Modeling Realistic Tanks and Artillery

An Illustrated Guide

Mike Ashey

KALMBACH
BOOKS

DEDICATION:
To my sons, Thomas Essa Ashey and Gregory Michael Ashey

 Published by Kalmbach Publishing Co., 21027 Crossroads Circle, Waukesha, WI 53187.

Printed in the United States of America

2000 01 02 03 04 05 06 07 08 10 9 8 7 6 5 4 3 2 1

Visit our website at
http://kalmbachbooks.com
Secure online ordering available

Publisher's Cataloging in Publication

(Provided by Quality Books, Inc.)

Ashey, Mike
Modeling realistic tanks and artillery : an illustrated guide / Mike Ashey. — 1st ed.
p. cm.
ISBN: 0-89024-371-9

1. Military miniatures. 2. Tanks (Military science)—Models. 3. Artillery—Models. I. Title.

U311.A69 2000 745.592'82
QBI00-36

Book design: Jonathan Johnson
Cover design: Kristi Ludwig

Cover photo: William Zuback

ACKNOWLEDGMENTS: Many people have helped me over the years and have in many ways, directly and indirectly, contributed to this book. Special thanks are due to Dr. Eric Lubot, Dr. Randy Frosstrum, and Dr. Mia Anderson of Bergen Community College, Paramus New Jersey; Dr. Ray McAllister, Professor Emeritus of Ocean Engineering, Florida Atlantic University; the Big Bend Scale Modelers Club; Glenn Johnson for his photography work; Scott Weller, Richard Boutin Sr., Daniel O'Keefe, and Bill Teehan of Tallahassee, Florida, Jim Hudson of Woodville, Florida, as well as Steven O'Connell, Jose T. Rodriguez, Keith Lehle, Owen R. Auer, Matthew W. Malogorski, Bob Skladany, Ed C. Kennedy, and Jim Hensley, for allowing me to showcase their models in this book; Terry Spohn of Kalmbach Publishing; Joel and Marian Swanson for their encouragement; my parents Joseph and Marie for instilling within me an engine that knows no rest; and, most important, special thanks are due to my sons, Thomas and Gregory, for giving me some space to get the book done, and my wife, Kelly, for helping me every step of the way.

Contents

Manufacturers and Suppliers

There are dozens of aftermarket and cottage industry suppliers that manufacture an endless list of detail sets. I have listed some of the main suppliers. But the best way to accumulate information on all these manufacturers is to send away for information whenever you see a manufacturer's advertisement or a new products release in *FineScale Modeler* or any other magazine.

Badger Airbrush
9128 West Belmount Ave.
Franklin Park, Il 60131
(Airbrush equipment)

Bare Metal Foil Company
P. O. Box 82
Farmington Hill, MI 48024
(Plastic polish, plastic scriber, small parts placer, and resin)

Borden Inc.
Department CP
Columbus, OH 43215
(White glue)

Cavalier Model Productions
Precision Parts Corporation
P. O. Box 305
Ogdensburg, NY 13669
(Resin detail sets)

Collectors Brass
P. O. Box 27191
Fresno, CA 93729
(Brass casting detail sets)

Creations Unlimited Hobby Products
2939 Montreat Dr. N.E.
Department 169
Grand Rapids, MI 49505
(Flex-I-File and Flex-I-File sanding sticks)

Detail Master
P. O. Box 1464
Sterling, VA 20167
(Engine wiring and detail sets)

Devcon Corporation
Danvers, MA 01923
(Two-part epoxy)

Dremel
4915 - 21st St.
Racine, WI 53406
(Drill bits, cutters, drill press, and drill press vise)

Eduard Precision Parts
P. O. Box 305
Ogdensburg, NY 13669
(Photoetched detail sets)

Evergreen Scale Models
18620-F 141 St. Avenue N.E.
Woodinville, WA 98072
(Plastic strips, shapes, rod, and sheeting)

Floquil Polly-S / Bondex International
206 Milvan Dr.
Weston, ON M9L 1Z9 Canada
(Paints, plastic prep, and paint/decal remover)

Grandt Line Products
1040 B Shary Ct.
Concord, CA 94518
(Bolt and rivet heads and detail parts)

Gunze Sangyo Detail Sets
Marco Polo Imports
550A S. Coralridge Pl.
City of Industry CA 91746
(Resin detail sets)

Jaguar Detail Sets
Marco Polo Imports
550A S. Coralridge Pl.
City of Industry CA 91746
(Resin detail sets)

K & S Engineering Co.
6917 W. 59th St.
Chicago, IL 60638
(Sandpaper, brass rod, and tubing)

Loctite Corporation
Cleveland, OH 44128
(Super glue)

Micro Mark
340-1513 Snyder Ave.
Berkeley Heights, NJ 07922
(Supplies and tools)

Microscale Industries, Inc.
P. O. Box 11950
Costa Mesa, CA 92627
(Decal setting solution)

Model Expo, Inc.
3850 N. 29th Terrace
P. O. Box 229140
Hollywood, FL 33022
(Supplies and tools)

NorthWest Short Line
P. O. Box 423
Seattle, WA 98111
(Chopper and True Sander)

Pacer Technology
9420 Santa Anita Ave.
Rancho Cucamonga, CA 91730
(Super glue accelerator)

Plastruct
1020 S. Wallace Pl.
City of Industry, CA 91748
(Plastic shapes and rod)

Scotch 3M Painter's Masking Tape
Box 33053
St. Paul, MN 55133
(Masking tape)

Small Parts, Inc.
13980 N.W. 58th Ct.
P. O. Box 4650
Miami Lakes, FL 33014
(Brass wire, tube, and stock)

Squadron Shop
1115 Crowley Dr.
Carrollton, TX, 75011
(Resin and plastic models, supplies, and photoetching)

Super Scale International, Inc.
Carson City, NV
(Decal setting solutions and white glue)

Testor Corporation
620 Buckbee St.
Rockford, IL 61104
(Paints, paintbrushes, putty, sandpaper, glue and glue applicators, sprue cutter, and airbrush equipment)

VLS Corporation
1011 Industrial Ct.
Moscow Mills, MO 63362
(Resin and photoetched detail sets and kits)

Waldron Model Products
P. O. Box 431
Merlin, OR 97532
(Waldron punch set)

X-acto
P. O. Box 5830
Statesville, NC 28677
(Blades and cutting tools)

Foreword

I wrote *Modeling Realistic Tanks and Artillery: An Illustrated Guide* in an effort to show modelers, with step-by-step pictures, how to build realistic scale models. Each chapter is designed to provide as much information as possible, and I have sequenced them to make it as easy as possible to locate specific information or review a technique. You'll learn how to build flawless models from start to finish; remove seams and fix fit problems; work with resin kits and fix resin casting problems with simple solutions; drill out gun barrels; work with brass photoetching; work with white-metal and brass castings; work with flexible track sections and individual track links; build up interiors; do basic scratchbuilding; and mask and paint, apply decals, and add weathering.

Since modelers create three-dimensional objects, and much of the modeling process requires visual interpretation, the majority of us in this hobby fall into the category of visual people. We respond well to visual stimuli. This book capitalizes on our visual sense by presenting pictures in place of words. While words are also valuable, it can take an entire page of text to describe just one picture. Since my motivation for writing this book was to pack as much information as I could into a limited number of pages, the best way to do this is to convey ideas, techniques, procedures, and successes through pictures.

If you really want to increase your modelbuilding skill level, learn new techniques, or learn how to scratchbuild, the best way is to build different kinds of models. If you love building tanks, that's great—but build a military vehicle or an airplane if you really want to explore new ways to build better models. You would be amazed at the things you will learn and the new tools you will discover. I discovered the Waldron Products punch tool when building up an aircraft cockpit console. This tool has thousands of uses for building tanks, aircraft, and ships. I discovered hundreds of uses for Bare Metal Foils plastic panel line scriber, which was designed for scribing aircraft panel lines. NorthWest Short Line's chopper and True Sander were designed for model railroaders, yet I use these tools all the time when I am scratchbuilding parts for tanks, military vehicles, ships, or aircraft. If I had not been exploring model railroading supplies, I would have never found these handy, simple tools.

I think I have accomplished what I set out to do. Always remember that the true nature of this hobby is creating scale models by exploring and expanding the boundaries of your abilities and imagination, and most of all, having fun.

Happy modeling!
Mike Ashey

1 The Basics – Modeling Tips and Techniques

Over the years I have discovered and rediscovered techniques for building scale models that have greatly advanced my modeling skills. I have condensed and refined them for presentation in this book so that anyone can use them with success. All the tips, techniques, and procedures presented in this book have been tested and proven over and over. Good luck—and remember that modeling is what you make it, so just have fun!

Tools. I like to keep my workbench neat and organized. Since armor kits can have a lot of parts and subassemblies, I use plastic storage boxes for organizing them. These covered bin organizers are usually found in the sewing sections of arts and crafts stores. I also make sure that I have ample supplies of new number 11 X-acto blades, an X-acto stencil knife, and single-edged razor blades.

Various grades of waterproof sandpaper ranging from 150 to 600 grit, various grits of sanding sticks, and a Flex-I-File are a must. A great place to find sanding sticks is the nail care section of your local drug or grocery store. Testors and K&S Engineering market waterproof sandpaper, and you can also find large sheets of it at automotive supply stores.

For basic scratchbuilding you will need a NorthWest Short Line plastic chopper and the same company's True Sander; a good supply of Evergreen rod, strip, and sheet stock; and various small diameters of brass rod. Drill bits are a must as well, for either a twist drill or a pin vise. Micro Mart and Model Expo are great supply sources for basic supplies, drill bits, and all kinds of tools.

For glue I like cyanoacrylate cement (commonly called "super glue"). You can get super glue that is as thin as water or as thick as pancake syrup. I also keep new tubes of Testors glue on hand. White glue is an excellent filler, and it works well for gluing clear parts. For fillers I use Testors gray putty and automotive putty. For weathering I use Polly-S weathering paints and pastel pencils. For decal application I use decal setting solution.

A Waldron Punch Tool is a good piece of equipment to have; it has a thousand and one uses. Bare Metal Foil's plastic panel scriber is also a fine tool for cutting and scribing plastic. Testors Metalizer paints are great for painting vinyl tracks. Testors Dullcote is good for sealing decals and pastel pencil dust, as well as toning down drybrushed silver paint. For masking I use Scotch 3M painter's masking tape. It can be cut into thin strips, it has excellent adhesion, and it will also stretch a little. Additional tools and supplies are mentioned throughout the text and photo captions.

Techniques. To have a successful experience building scale models, it's important to study the instructions first and become familiar with all the parts and assembly sequences. I decide how I will assemble the kit, and I use the instructions only as a guide. This is an important point because the assembly sequence that you should follow will depend on the color scheme, details you may want to add or changes you may want to make to the kit, and how you want to deal with any problems you have identified. At this time, I also assemble reference material that I want to use for detailing and painting.

When removing parts from "trees" (sprues), be careful how you cut them off, because this is the first step in seam removal. If you snap off the parts, you will most likely damage them, which means repair work either to the part or extra work to fill a seam. I spend a lot of time test-fitting parts to find out where the fit problems are. I also tape together the main parts to get a feel for the size of the model and how it will look when completed. Every model is different. No matter who the manufacturer is, each model has its unique fit problems and minor flaws. Whatever the problem, however, there is always a simple solution.

Check each part for a mold line and carefully scrape off this line with the tip of a sharp number 11 X-acto blade. For parts that are glued together, check the gluing surfaces—you will sometimes find small injection marks that must be flattened. Check the fit of these assemblies, because the locating pins are not always correct. Sometimes you can get a better fit by removing the pins and running the part halves across a stationary piece of sandpaper to flatten the gluing surfaces.

To glue parts together I secure them with small lengths of masking tape and then apply super glue along the seam line with a thin wire applicator between the masking tape areas. The capillary action of the super glue draws the glue down into the seam surface area, resulting in a very strong bond. Once the glue is dry, I remove the tape and apply glue to the remaining areas. I then carefully scrape and sand the seam line. To check for flaws in my seams, I apply silver paint to highlight them. I apply more super glue to these areas, scrape and sand them, and then remove the remaining silver paint with Polly-S Paint and Decal Remover. Be sure to wash the parts to remove any paint remover residue. Be careful when using Polly-S paint remover on resin—it will eat into the resin if you do not wash it off immediately after you remove the unwanted paint.

Sometimes the surfaces of parts have small, round indentations. These are from the injection-molding process, and there are several ways to get rid of them. If the round indentation is deep, you can fill it by punching out a thin plastic disk using a Waldron punch tool, gluing it in place, and then sanding it smooth. You can carefully scrape off shallow indentations using an X-acto stencil knife and then sand the surface smooth.

When you check the fit of subassemblies or when you glue parts to subassemblies that are already painted, you will sometimes find voids. The secret to dealing with these is to fill them with white glue applied with a thin wire applicator and then remove the excess with a damp cotton swab. Sometimes several applications are necessary to completely fill a void. Since the glue dries clear, it transmits the surrounding colors right through so well that you may not have to paint it.

Painting tips. I decide what color scheme I want and I assemble the paints and mix the colors, test them, and make any necessary adjustments. Mixing the colors is also the first step in weathering. I always lighten any color paint that I am using, and I also mix small quantities of even lighter shades. Lightening the paint color ensures that the finished color will have a scale effect. As an example, olive drab is the color that would be applied to an actual tank. Since your model is smaller, the olive drab will appear too dark, so you will need to lighten it just a little. After I have finished applying the base color, I apply additional coats of the lighter color to simulate the fading effect the sun can have on horizontal surfaces and other areas that get direct exposure.

There are several simple steps in painting that are an absolute must if you want to get good results. First, you have to ensure that all parts are clean and free of sanding dust and residue. Give all the parts a cleaning with Ivory soap and warm water or Polly-S surface preparation cleaner. Give all metal castings and photoetched parts a final cleaning in mineral spirits or some other type of enamel-based thinner.

To get a quality finish, or to mix colors for various shades and fading effects, you cannot get by without an airbrush. I like to use a Badger 200 single-action airbrush. Since temperature and, to a large degree, humidity affect airbrushing, you should do all your painting in an environment where the temperature is moderate and the humidity is low. Generally, temperatures between 65 and 75 degrees are good, and humidity of no more than 55 to 60 percent is acceptable. The thinner that you use is also important; I always use the manufacturer's recommended thinners.

To get the most from each bottle of paint, drop a few copper BBs into the bottle to help mix up the sticky paint on the bottom. Thinning paint for airbrushing can sometimes be a trial-and-error process. Generally, I use a mixture ranging from 4 parts paint to 1 part thinner to 3 parts paint to 1 part thinner at 15 to 20 pounds of air pressure. I have tried a lot of different air supplies, and the easiest and quietest by far is a CO_2 bottle with a pressure regulator. While the setup cost is a little more than for a small compressor, the air is always dry, the air flow is consistent and adjustable, and it is quiet. A bottle typically lasts for two years or more. When you're painting a surface that has a lot of protruding parts, like a tank turret, I suggest that you use low air pressure (15psi or less). Low air pressure will help prevent the dusting or "orange-peel" effect that can occur behind a protruding part if you spray it from the front with high pressure. Paint particles dry before they land on the area behind a protruding part, causing the dusting effect.

When you are ready to paint small parts, secure them with masking tape on lengths of balsa or stiff cardboard sections so you won't be handling them. Give all parts a coat of primer first. The only time you shouldn't use primer is when you are using Testors Metalizer paints. Primer provides a good adhesion surface for paint, and the primer color will also highlight any flaws, cracks, seams, and scratches you may have missed. If you do find any areas that need additional finishing work, let the primer dry for a few days, fix the problem areas, and then sand the surrounding primer using a minimum of 600-grit sandpaper. To blend in the unpainted and primer-painted areas, give the unpainted area a coat of primer first and then give the entire area a complete coat.

Don't rush the finish coats of paint. Let the paint dry for at least 24 hours before you handle the model or apply a second coat. Gloss paint takes longer to dry, and it may be several days before a gloss paint color like white really dries. If the surface smells like fresh paint, it's still drying.

Keeping your airbrush clean is also very important. I run thinner through the airbrush between colors and clean the tip. I also check the tip during a spraying session, because sometimes paint will build up around the edge and then splatter onto the surface. When each session is finished, I disassemble the airbrush and clean it using tissues, thinner, and pipe cleaners. To make enamel-based paint flow more smoothly, I sometimes warm the paint on a coffee cup warmer. I always keep the lid open a crack during the warming process so that pressure will not build up inside the bottle. The paint only has to be on the warm side to get it to flow smoothly.

If you are using spray paint, let the can rest upside down for a few hours, and then give it a good shaking. Spray paint can also be warmed by setting the can in warm water for a few minutes. Always test the spray paint before using it and clean the spray tip when you are done. To remove excess paint from the tip, hold the can upside down and spray until no paint comes out. If you are using a hand brush, thin the paint slightly so it will flow better. Paintbrush selection is also important. I use only natural-hair brushes that keep their shape. Clean your brushes after every use, and run them under hot water to reshape them.

Decals. The secret to decal application is to apply them to a gloss surface and use decal setting solution to get decals to snuggle up around surface details. If you used gloss paints you are ready to apply the decals; if you used a flat paint, you'll have to add a clear gloss finish. I have had a lot of success using clear gloss polyurethane paint, the kind you find in hardware or home supply stores. If you use polyurethane paint, you only need to thin it slightly to get it to work in an airbrush. Gloss surfaces will also allow you to slide the decal around to position it correctly.

Always cut decals free from their sheets using a sharp knife and a ruler. Cut off as much of the clear film as possible. For small decals, cut around the perimeter as close to the edge as possible. For large numbers or letters, cut out each one individually, remove all the clear film, and then apply them one at a time. The trick is to ensure that they are lined up correctly. For curved areas like circles, use a series of tangential cuts to trim away the excess film. For straight cuts, use a metal ruler as a guide.

If the decal starts to dry while you are positioning it, moisten it with some water applied with a cotton swab. Let the decal dry when positioned, then apply coats of setting solution to the surface using a cotton swab. After a few coats the decal will soften and snuggle up around the raised detail. When you have finished applying all the decals, clean up the surrounding surfaces to remove water stains, and then add a coat of Testors Dullcote to protect the decals.

Weathering. The final step is weathering, but you have to be very careful not to overdo it. As I explained earlier, the first step in weathering is to ensure that the colors you use to paint the model are lighter than the actual colors. Once the decals are applied, you will have new-looking decals and faded paint, which will not look right. To fade the decals, apply a dilute mixture of water-based flat white or light gray paint to the surface of the model. Apply the diluted paint lightly and let it dry before applying more coats. Be sure to check your work carefully, as it is very easy to overdo this technique. Use water-based paint because you will be spraying a lot of thinner onto the painted surface and the water will not affect the surface. If you were to use enamel paints, the thinner would smear the surface paint. Note that the diluted paint will also lighten the overall color of the base paint.

Before I apply coats of paint to simulate dirt, mud, and dust, I drybrush with Testors silver to highlight all the edges, surface detail, and areas where the paint would be worn away, exposing the bare metal. To tone down the silver paint, I apply a coat of Testors Dullcote. Be sure to drybrush all the subassemblies to highlight those parts too.

Once you have lightened the decals and drybrushed, you are ready to apply dilute coats of mud- and dirt-colored paint. Polly-S has an excellent set of weathering paint colors. Apply around the wheels, bogies, sprockets, rollers, and tracks, and lower areas of the hull of the tank or vehicle frame that would be in close contact with the ground. I also use pastel pencil dust to simulate exhaust stains around mufflers and engine gratings as well as powder residue on the ends of gun barrels.

Use a set of thin-bladed wire cutters or a plastic sprue cutter to remove the parts from their trees, and be sure to leave some of the excess plastic stem on the part.

For small parts, set the tree on an elevated surface like a small hardwood block and cut the stems with a number 11 X-acto blade. Here again, leave some of the stem attached to the part.

For small, delicate parts, cut the trees around the part and then trim it on a hardwood block.

It is much easier to trim off the excess plastic from the stem without having to maneuver the blade around other parts still attached to the tree. Here again, do your trim work on a hardwood block.

All along this framing, there are numerous injection marks, which you'll have to file and sand smooth. Identifying these problems early helps you plan your construction sequence.

The injection marks on the inner surfaces of these parts will be hard to get to. If the marks will not show on the finished model, don't mess with them. As an example, if you plan to keep the hatches closed, there will be no need to remove the marks from inside the hatch.

The sink marks on these road wheel suspension arms is a good example of how sink marks can mar the surface of a part. Fortunately, since the road wheels will hide them, there is no need to fix them.

tips

- Do not snap parts off trees.
- Always use a sharp blade to remove parts.
- Leave excess plastic on the parts. You can then carefully trim it off on a raised surface.

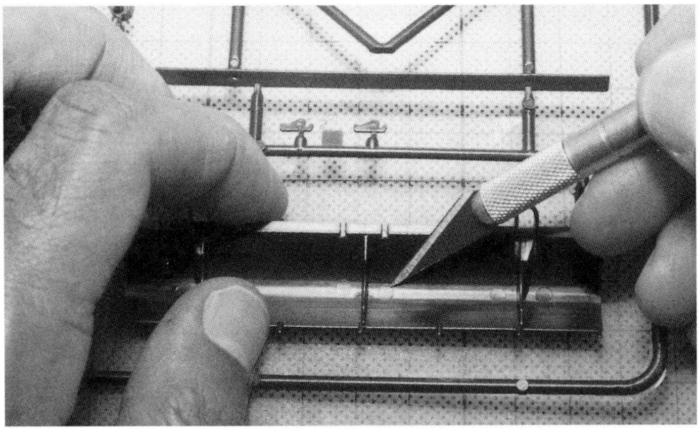

Since the face of this shovel will be one of the eye-catchers of the overall model, you'll have to remove the injection marks on the inside curve. This will require creativity.

Plastic bin organizers found in arts and craft stores are perfect for storing and organizing all the parts as you remove them from the trees and clean them up.

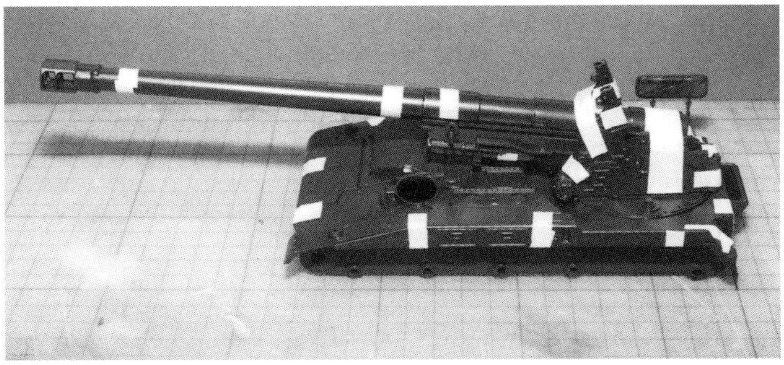

Above are three good examples of fitting together the main components of these different types of models. Building up the main components using strips of masking tape is the best way to identify fit problems and voids. It also gives you an opportunity to see how the model will look when completed and get you thinking about paint schemes and weathering.

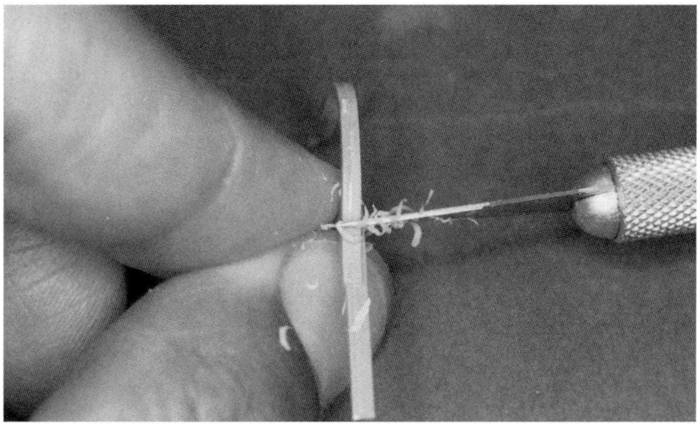

After you remove each part from its tree and clean and cut off the excess plastic, carefully scrape the mold lines.The tip of a number 11 X-acto blade is great for this.

Sometimes part halves fit together better without their alignment pins. Flatten an engine block half's gluing surface by rubbing it gently across a piece of stationary sandpaper.

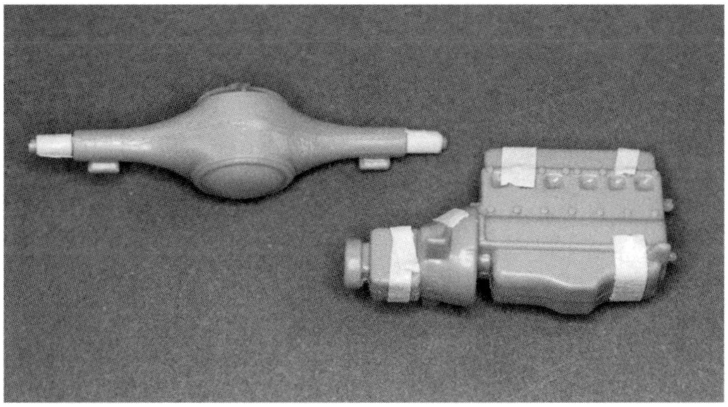

After flattening the gluing surfaces of the two parts, tape them together. They are now ready for gluing.

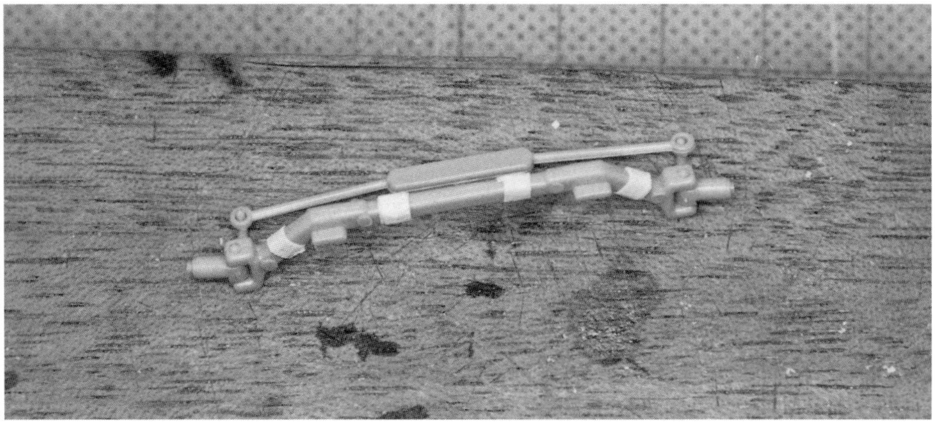

Here is a good example of taping two part halves together and also adding the necessary components to ensure that everything will be aligned correctly.

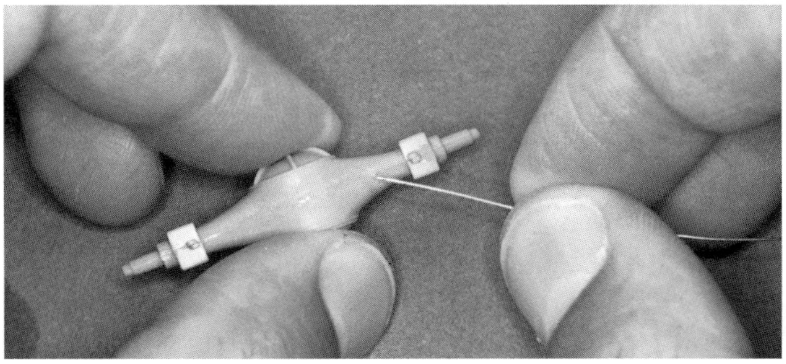

After you have taped the parts, the next step is to run a bead of super glue along the seam line using a thin wire as an applicator. To do this squeeze a puddle of super glue onto a piece of paper, dip the wire into the puddle, and then apply it along the seam line. Once the glue is dry, remove the tape and finish the job.

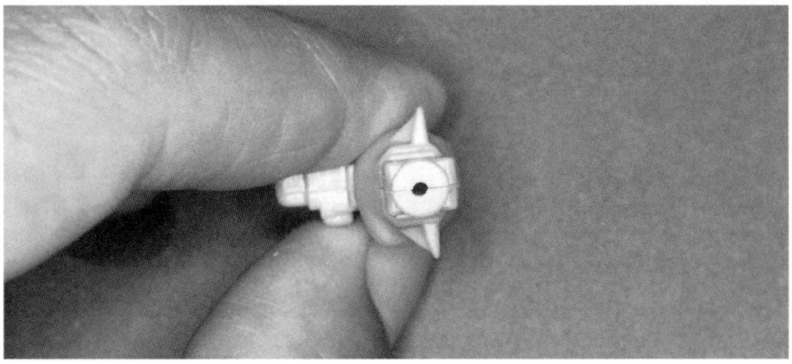

Here's a good example of how alignment pins would have severely misaligned these two part halves. If you used the alignment pins, the hole would have been perfectly round, but the rest of the part would have been misaligned. If you remove the alignment pins, the part halves line up perfectly. All that's necessary is to drill a slightly larger hole.

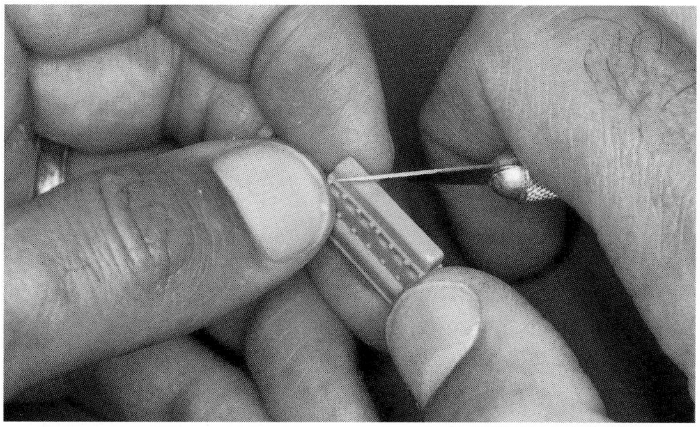

Once the super glue is completely dry, use the outer edge of a number 11 X-acto blade held at a slight angle and carefully scrape the super glue flat. The last step in dealing with this seam is to smooth out the glue with fine-grit sandpaper.

Here's a good example of the discoloration that super glue can cause on plastic. Since this part is a box shape, instead of scraping and then sanding, run the part across a stationary piece of sandpaper to smooth out the glue and blend it into the plastic.

Once you have completed your seam work, check to make sure you haven't missed anything. Testors silver paint works well as a crack detector.

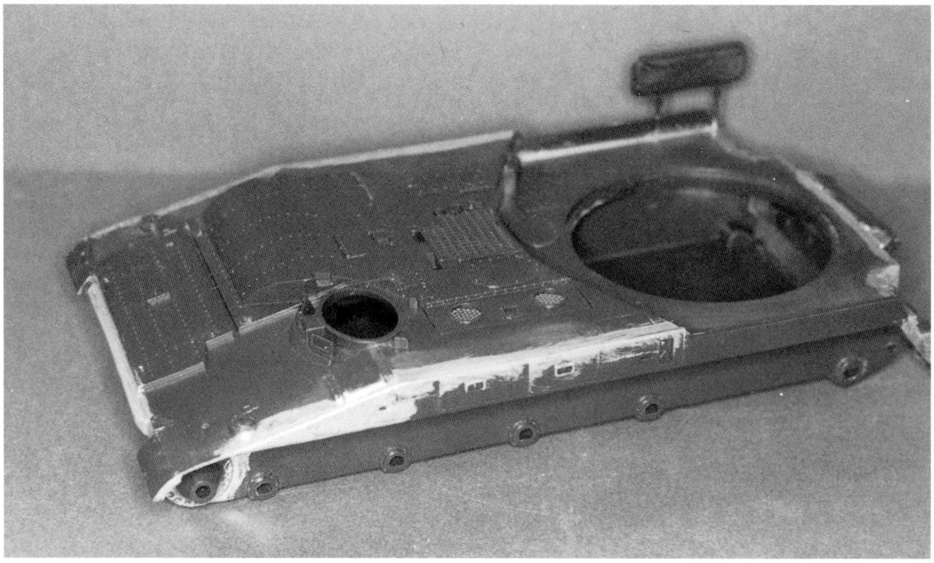

Here Testors silver paint is used to check all the seam work on this self-propelled artillery chassis. If you find any additional flaws, add more super glue, scrape and sand smooth, and then remove the silver paint with Polly-S Paint and Decal Remover. As a last step give the plastic a soap-and-water wash to remove any paint remover residue.

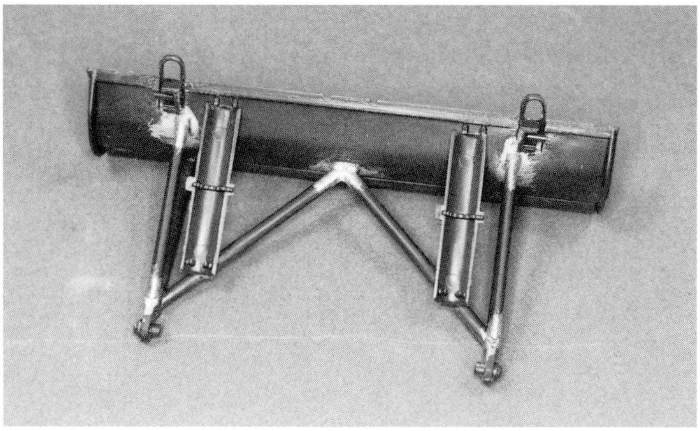

Add super glue at the connection points to simulate the weld, and use silver paint to check the simulated weld seams. After the Testors paint highlights additional areas that need glue, apply the additional coats of super glue.

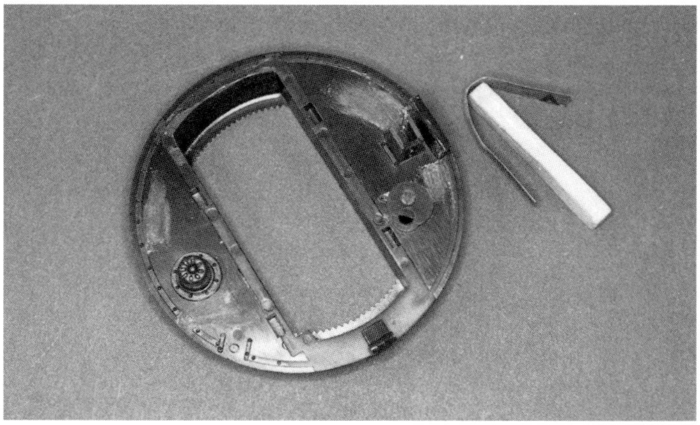

The injection marks on the surface of this part were in difficult-to-reach locations. Small strips of sandpaper wrapped along a length of balsa forms a small sanding tip that allows you to get into these tight locations.

A Waldron punch tool is great for making disks that can be glued into injection-mark openings. Glue the disks in place with super glue. Be sure to apply a small amount of super glue to the outside diameter of the disk so that it's completely sealed.

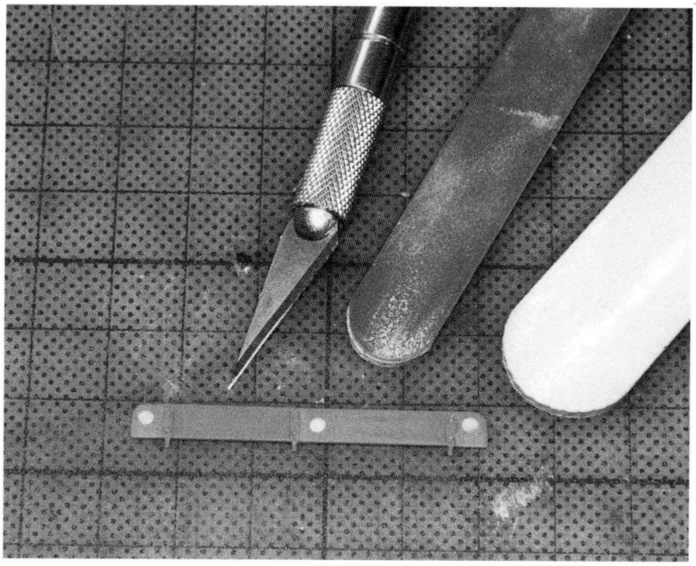

The next step is to sand and scrape the disks and super glue flat and blend it into the surface of the plastic.

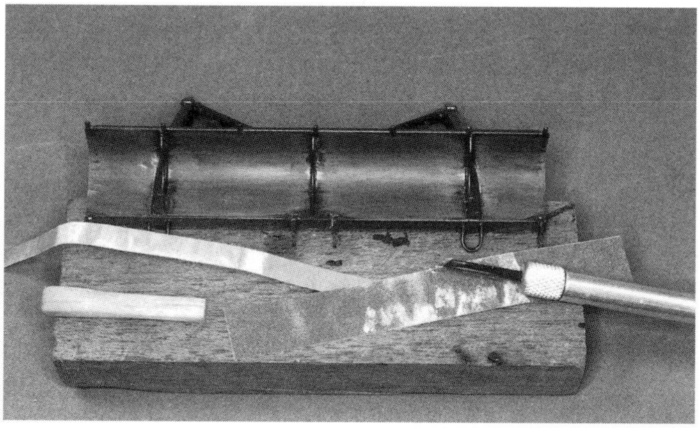

To get rid of the injection marks on the inside area of this shovel, I used thin strips of sandpaper wrapped around lengths of balsa and scraped with an X-acto stencil knife.

tips

- Long, thin strips of balsa make great sanding sticks for getting into tight places.
- Use Testors silver paint to check all your seam work.
- A Waldron punch tool has many uses in building tanks, artillery, and ground vehicles.
- A Testors stencil knife is great for scraping in tight places.

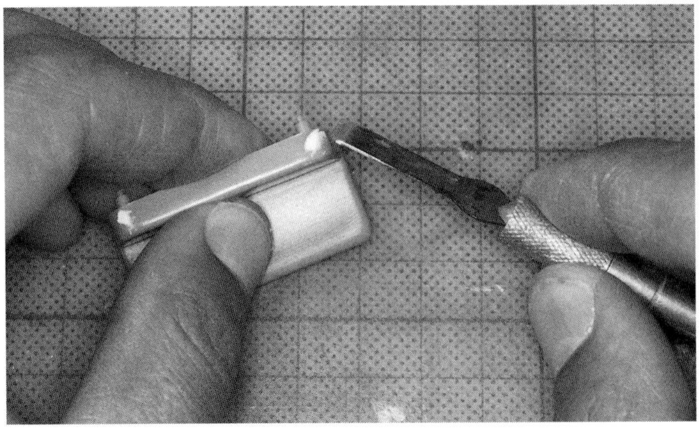

To fill the sink marks at the corners of this seat, apply small amounts of Testors putty with a flat-faced blade.

Then sand the putty smooth and blend it into the plastic.

The back sides of this part have deep injection marks. They are easy to fill with small lengths of Evergreen plastic rod.

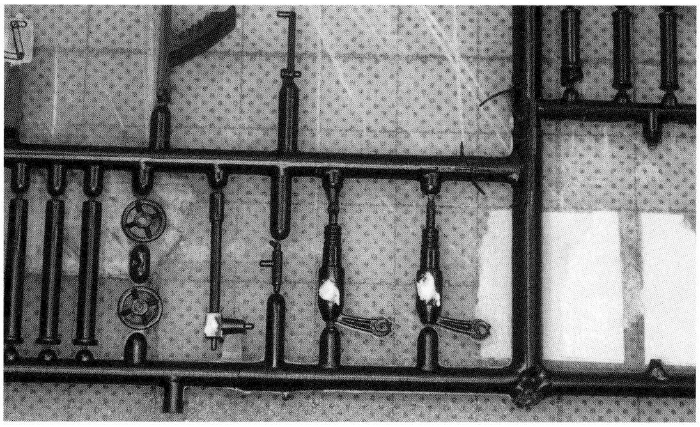

There were a lot of sink marks on all parts on this tree when I inspected it. To ensure that I caught all the sinks marks, I simply applied the putty while the parts were still attached to the tree.

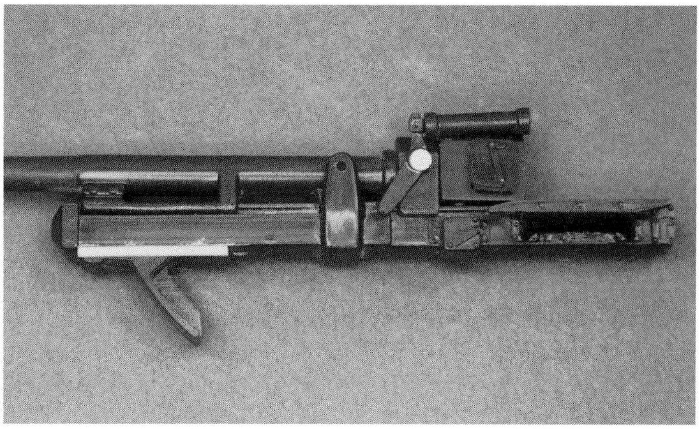

Sometimes hiding seam lines or filling injection marks would be difficult because filling them would damage the surrounding detail. In these two cases I used a small length of Evergreen plastic half round to cover the seam and a disk punched out by a Waldron punch tool to cover the injection mark.

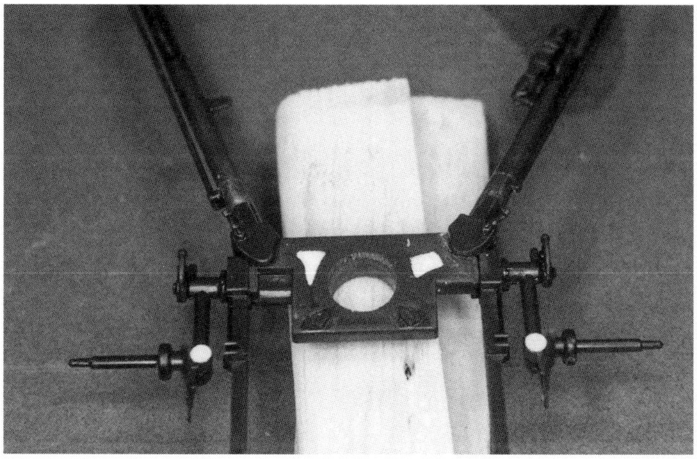

Here's an example of using putty to fill large sink marks and using small disks punched out by a Waldron punch tool to cover small injection marks. Different problems require different solutions.

tips

- Testors putty is great for filling small sink marks.
- Evergreen plastic rod and strips work well for filling voids or hiding flaws.

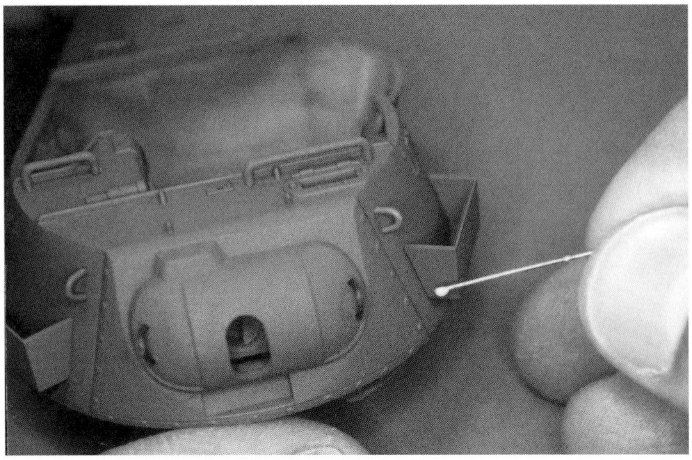

To fill the seam area between the turret and the external storage bin, I use white glue applied with a thin wire applicator.

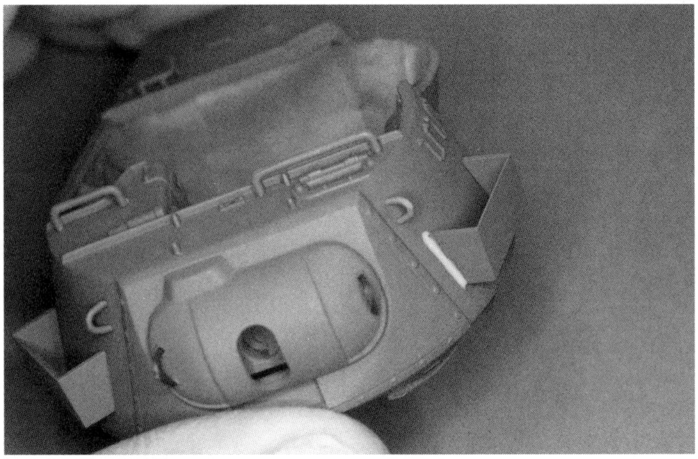

Here the white glue application is complete, and the next step is to smooth out the glue. Since skin forms rapidly on the surface of white glue, you have to apply the glue a little at a time.

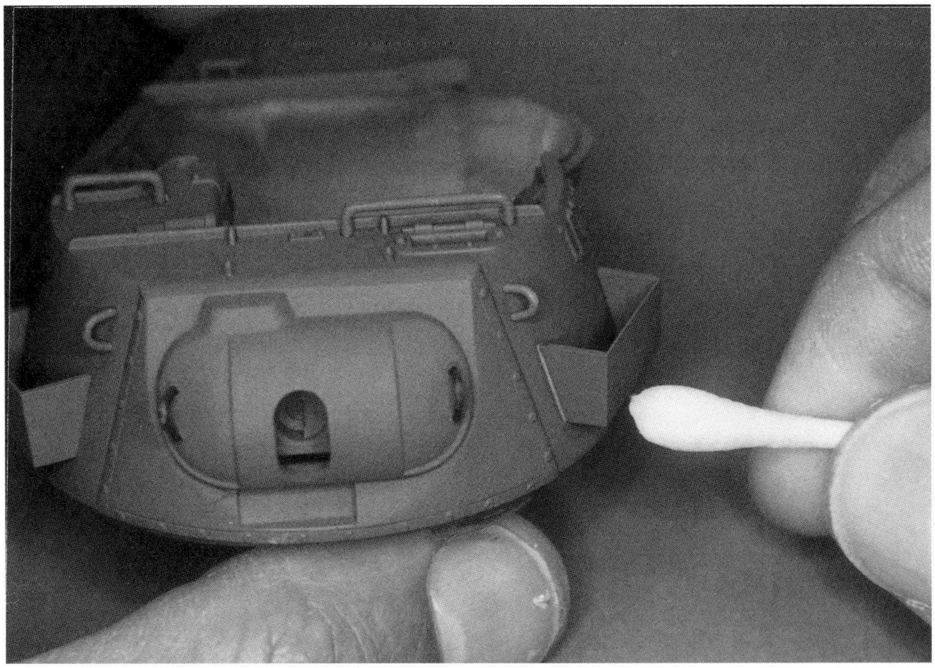

The next step is to moisten a cotton swab in your mouth with a twisting motion so that there is no fuzz on the swab, and then run it along the surface of the white glue to contour the glue and remove the excess.

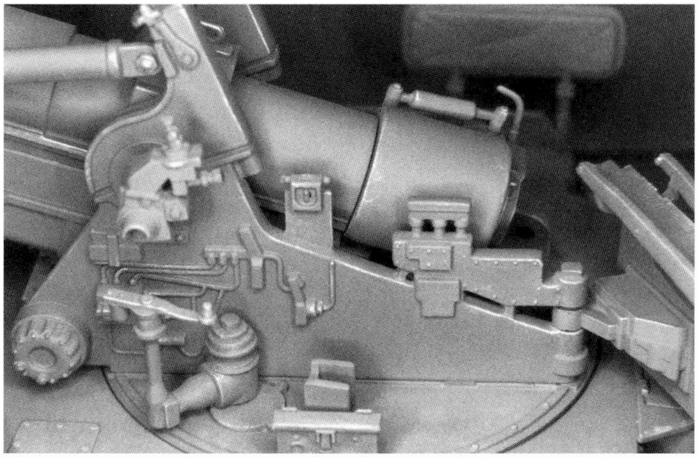

The void between the surface plate and carriage of this self-propelled gun would be impossible to fill with putty. Several applications of white glue solve this problem. Another advantage to using white glue is that it dries clear. It transmits the paint color of the surface that it is attached to so well that in many cases you don't even have to paint the glue.

White glue is also an excellent adhesive for clear parts because it will not mar the plastic. Clean excess glue off clear plastic with a moistened cotton swab. White glue is also an excellent medium to use as a filler in areas where there are voids between the clear part and the surrounding plastic.

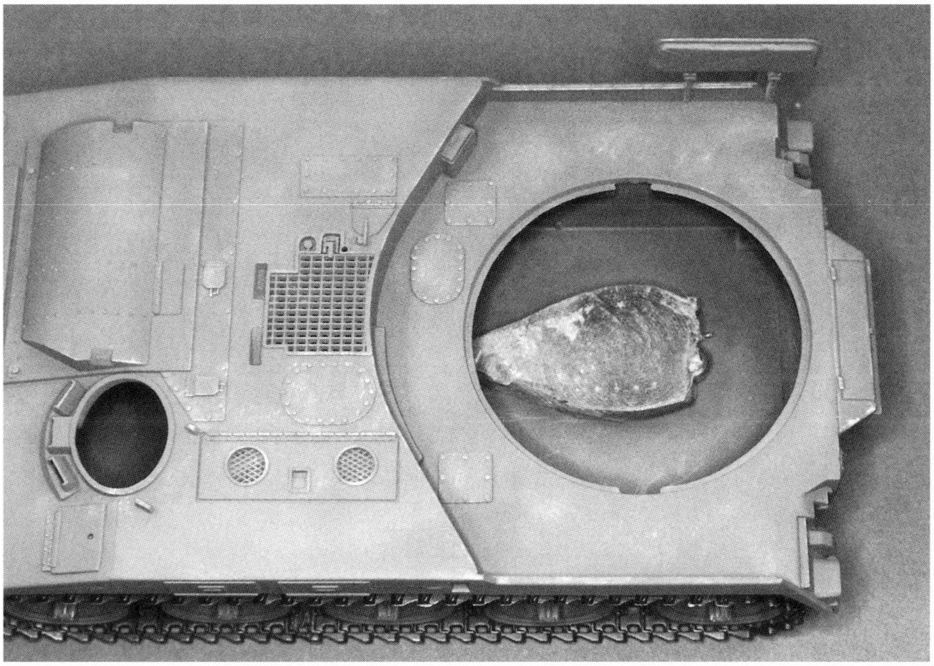

Adding weight to track vehicles will make the road wheels sit on the tracks instead of appearing to float over them. Do this by adding mashed lead fishing weights to the inside hull area.

To help mix the paint, add a few copper BBs to the paint container.

The best way to mix paint for airbrushing is to use two paint bottles of the same type and volume—one for the paint and one for the thinner. The mixture will be 3 parts paint and 1 part thinner.

I usually mix several shades of the primary colors I plan to use as well as the weathering washes. It is not unusual for me to end up with 10 or 12 bottles of mixed paint per model.

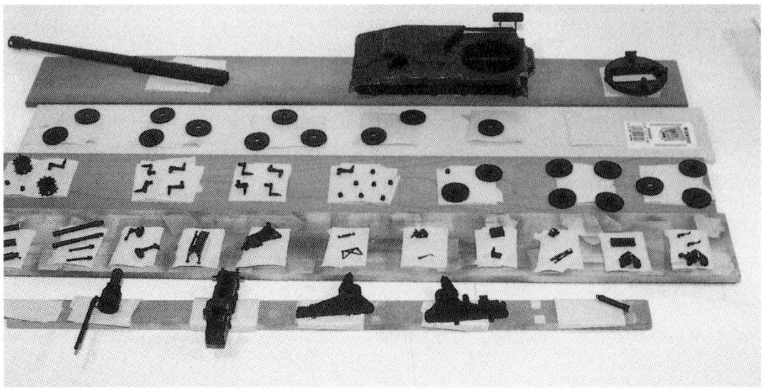

Use strips of wood or stiff cardboard and masking tape to attach all parts to be painted. Once the paint has dried, flip the parts and paint the other side.

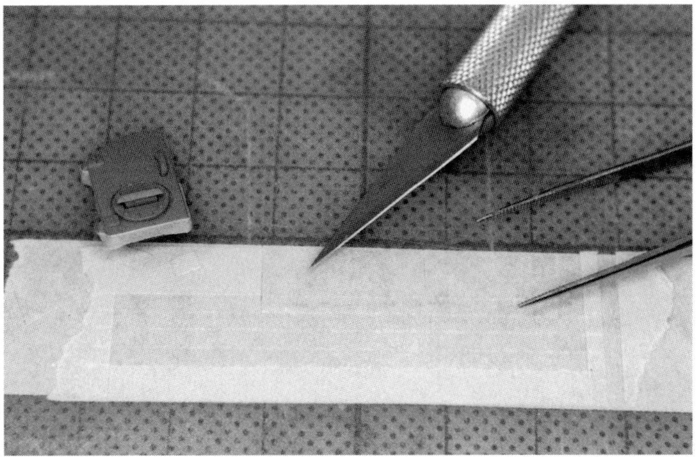

Cut small strips of masking tape by using two layers of tape. You usually don't use the bottom layer. Use a small metal ruler and a sharp number 11 X-acto blade to cut the strips.

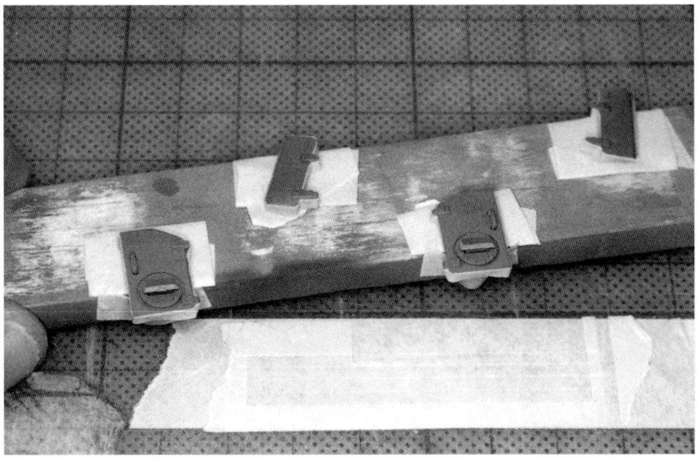

After you mask the edges of these hatch covers, they are ready for the top coat of olive drab.

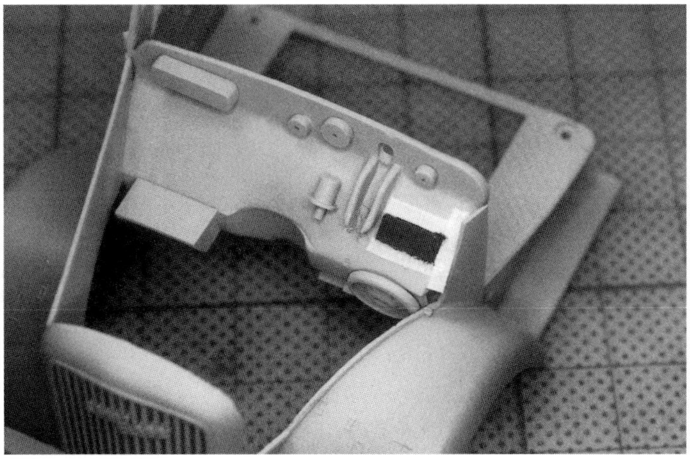

Apply small strips of masking tape around the perimeter of the fuse box and then brush-paint the part. As long as you apply small amounts of paint with a brush around the edges of the masking tape you shouldn't get any bleeding of the paint under the tape.

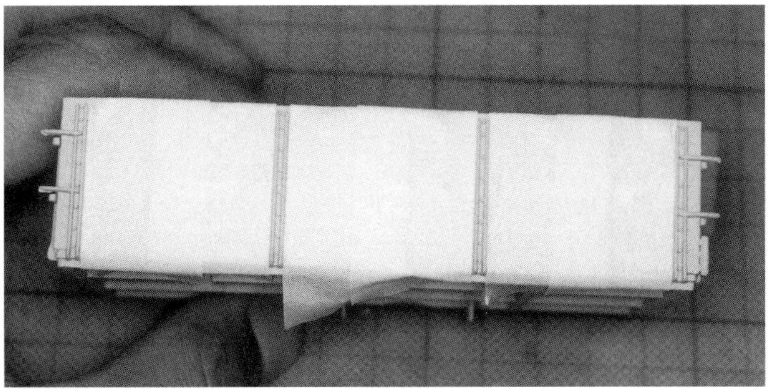

To paint the metal strips of this wood truck bed, position large strips of masking tape. Paint the strips a slightly darker shade of desert sand to differentiate them from the lighter wood.

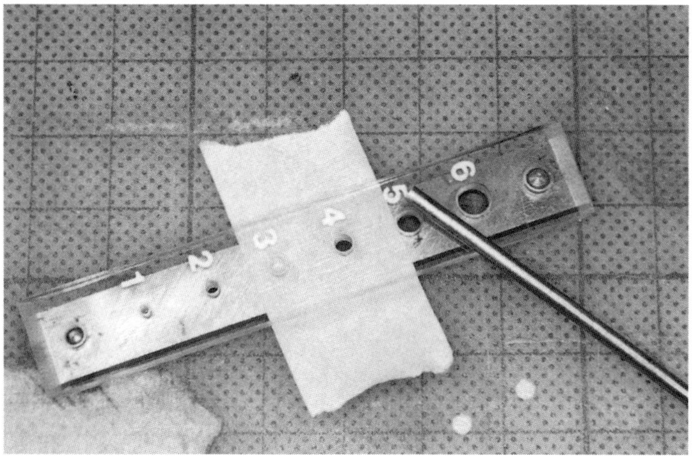

A Waldron punch tool is great for punching curves in masking tape.

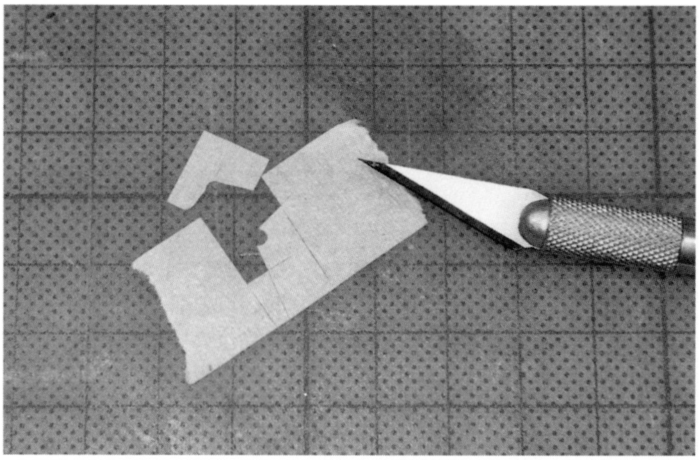

After punching out the curved area, cut it out. It is now ready for installation around the curved part.

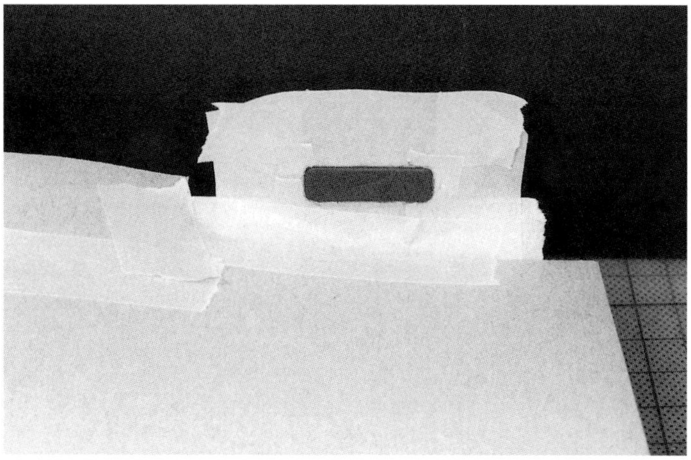

Attach the masking tape with the curved edges. The completely masked part is ready for painting. Note the white sheet of paper covering the tank's body.

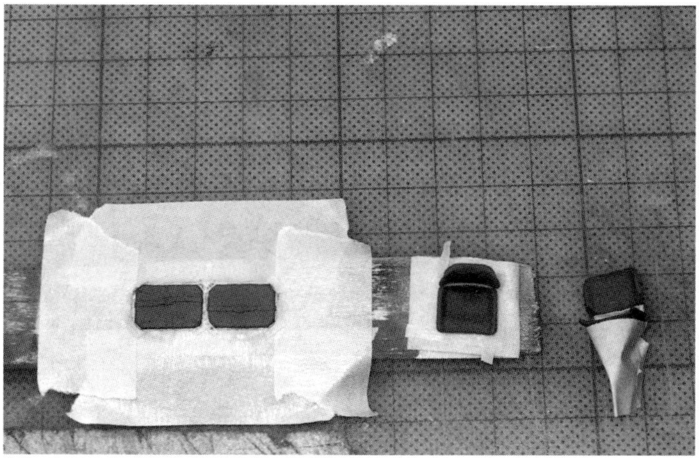

Here are some more examples of masking using combinations of large and small strips and lengths with curves punched in them using a Waldron punch tool.

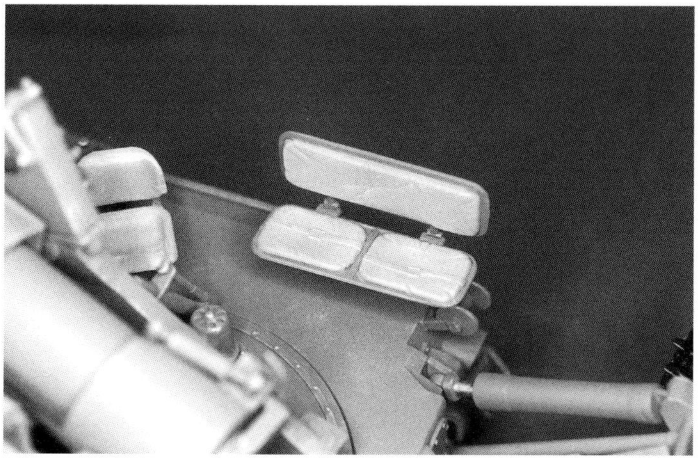

Here are the finished painted seats. Note the light canvas color to differentiate the cushions from the metal and the fine demarcations between the light cushion color and the dark metal color.

Fill a tank interior with tissue paper and then cover it with masking tape to ensure that airbrushing will not bleed into the interior.

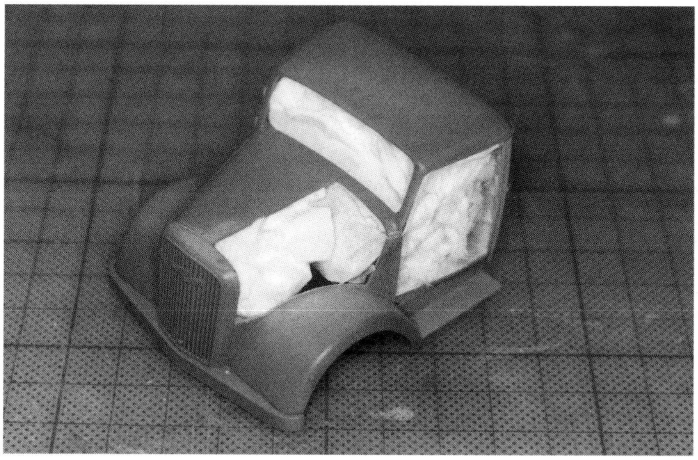

Here's another example of using tissue to protect interior colors. Mask the edges of the cab and then stuff the inside of the cab completely with tissue.

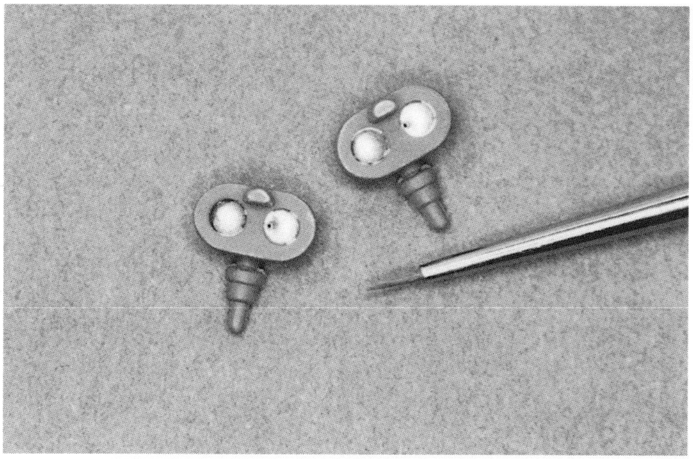

To pick out tiny details like headlights, use a small detail brush with a stiff tip.

tips

- A Waldron punch tool works well for punching out curved surfaces on masking tape.
- Tissue paper is great for filling the large interior areas of tanks or vehicles when painting.
- Always prime plastic prior to painting the finish colors.

The first step in drybrushing is to remove almost all of the paint from the brush.

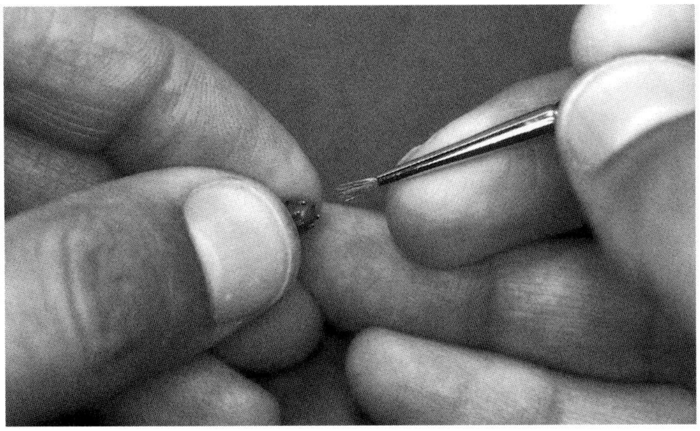

After removing almost all the paint, run the brush lightly over surface detail. The remaining paint particles on the brush will cling to the raised detail.

Here are three examples of drybrushing. Note how the bolt heads on the road wheel and hub stand out. The face and teeth of the dry sprocket received a liberal coat of silver to simulate the heavy wear of the track on the sprocket.

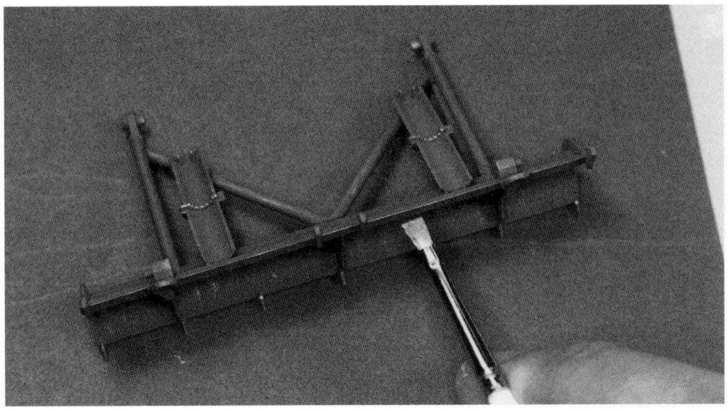

To make parts stand out, drybrush around all edges and corners.

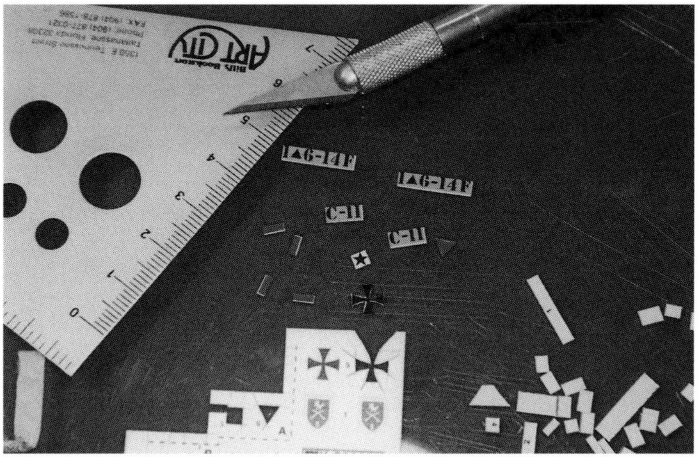

To reduce the silvering effect of decals, remove as much of the clear film on the decals as possible. Cut all your decals on a glass or Plexiglas plate. Always use a straightedge and sharp blade.

Remove all the clear film from large decals, and remove as much of the clear film as possible from smaller decals.

The secret to good decal application is to apply the decal to a gloss surface. I use clear gloss polyurethane applied with an airbrush. The gloss surface helps eliminate the silvering of decals and allows you to slide the decal around to position it correctly. Here the top rear turret has been glossed and is ready for decal application.

When you're ready to apply the decals, do them one at a time. Dip each decal in warm water for about 10 to 15 seconds and then remove it.

The warm water will loosen the decal's glue, and within 30 seconds or so the decal will slide off its backing. Move the decal slightly off its backing, grab the backing with tweezers, lay the decal over the approximate location, and then slide it onto the surface using a cotton swab soaked in decal setting solution.

Here's another decal being applied. Note that the clear film on the inside of the number four has been removed from this decal. Because of this hollow area in the center, you have to be careful that you don't fold the decal over on itself when you're sliding the decal off the backing.

The tank identification (F.4) was one decal with a lot of clear film. Instead of taking the chance of having a small portion of that clear film silver even on a gloss surface, cut out the individual decals, and apply them separately. The trick here is to make sure that the decals line up correctly.

This tank name was all one decal applied to a high-gloss surface and then coated with Testors Dullcote. This decal was also weathered with a dilute wash of mud. Note that there is no silvering of the clear film.

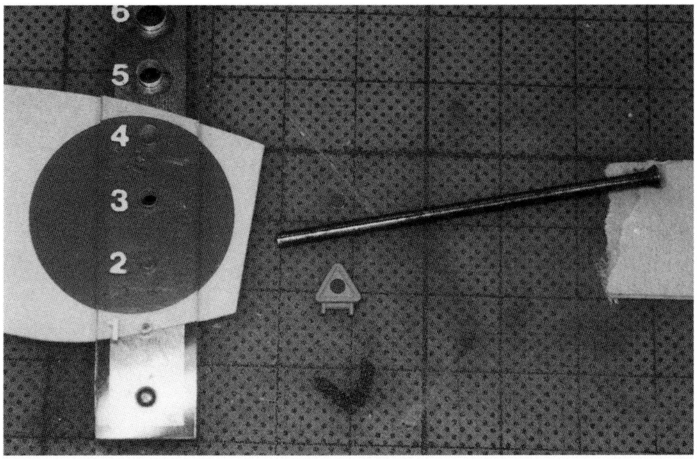

Sometimes you can use decals in place of paint. To get a small red dot like this, use a Waldron punch tool to punch one out from a spare aircraft decal sheet.

The decals on this M-18 Hellcat are now ready to be lightened so that they appear to be faded.

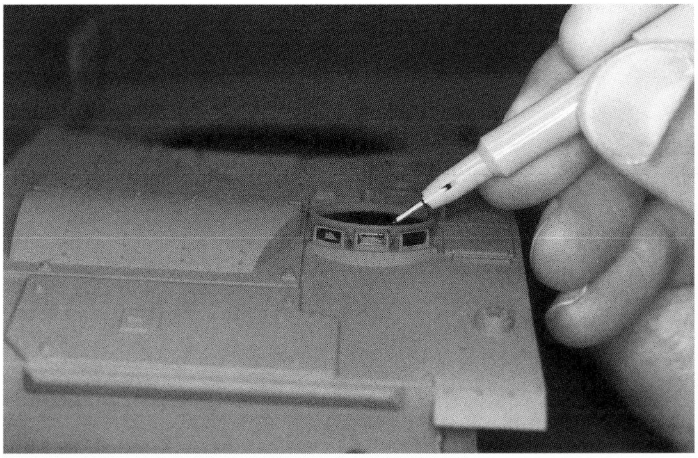

The easy way to paint periscope faces is to use a fine-tip disposable drawing pen.

Light coats of olive drab paint on the horizontal surfaces of this armored car in combination with careful drybrushing along the edges and in areas that get excessive wear help create the overall effect of a used and weathered vehicle.

tips

- Remove as much clear film from the decal as you can.
- Remember to gloss the surface where the decal will be applied.
- Use damp cotton swabs to help position the decal.

Here's a good example of using a dilute coat of umber colored paint in combination with drybrushing to help simulate heavy wear and use. Model by Richard Boutin, Sr.

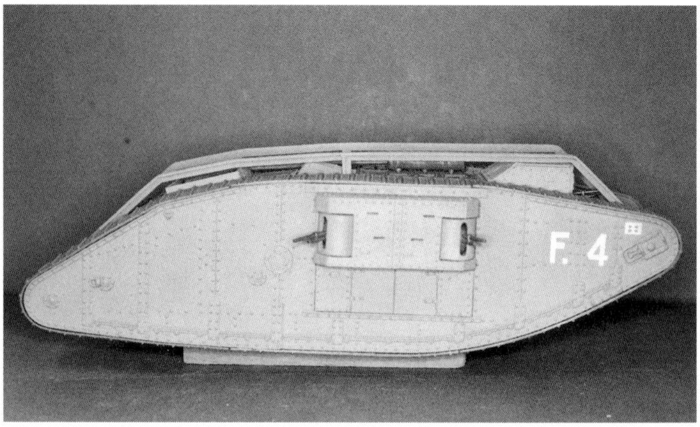

Apply dilute coats of mud, dirt, and rust to the lower areas of this British female tank to give it a weathered appearance. What you want to achieve is a subtle effect and hints of dirt and mud. It is better to underdo weathering than to overdo it.

To simulate splashed mud, as on the door of this Opel Blitz, apply an excessive coating of dilute mud so that it will smear downward.

These tires have a coating of dust so that they almost appear to be a lighter shade of dark gray rather than flat black. Note also the subtle mud splattering on the decal.

A good example of weathering on a towed self-propelled gun. In this case add all the applications of paint and dilute weathering coats first and then drybrush the edges. To tone down the silver paint color, the model received an overcoat of Testors Dullcote.

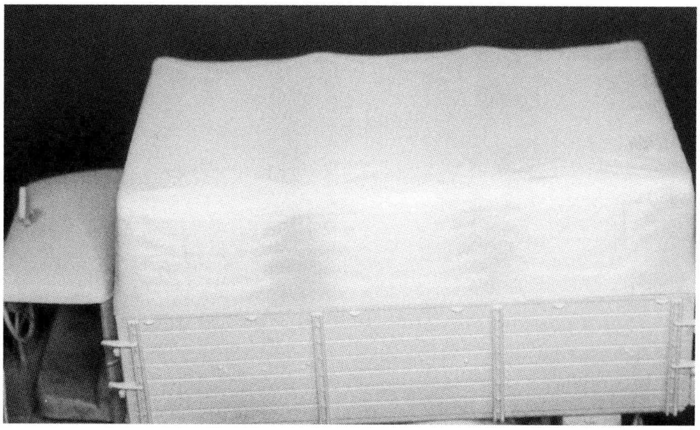

Here is an example of a subtle weathering effect on canvas. Note the four dark streaks across the canvas. These dark streaks simulate the bleeding and wear that the metal frames under the canvas would normally have on the outer surface of the canvas.

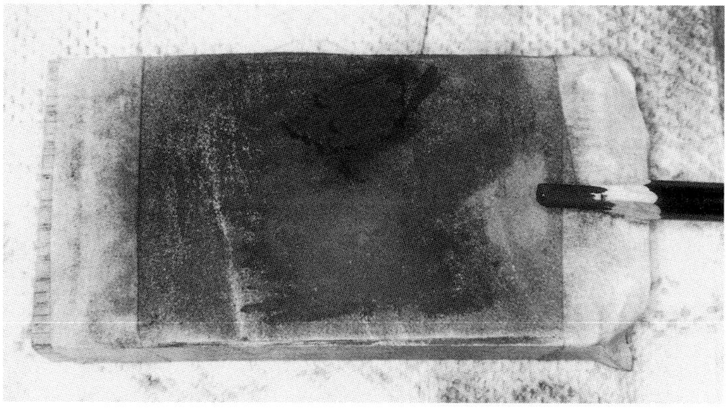

The first step in using pastel pencil dust as a weathering agent is to scrape the pencil onto sandpaper.

Use a soft flat brush to pick up pastel dust.

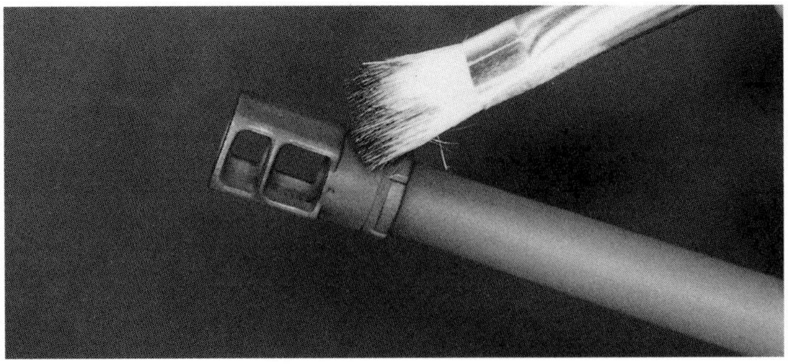

Apply black pastel dust to the muzzle brake area of a barrel to simulate smoke and powder burns.

Apply pastel dust to the engine cover areas and exhausts on this M-18 Hellcat. Additional pastel dust applied on some of the surface areas of the metal simulates discoloration of the surface paint from the heat of the engine.

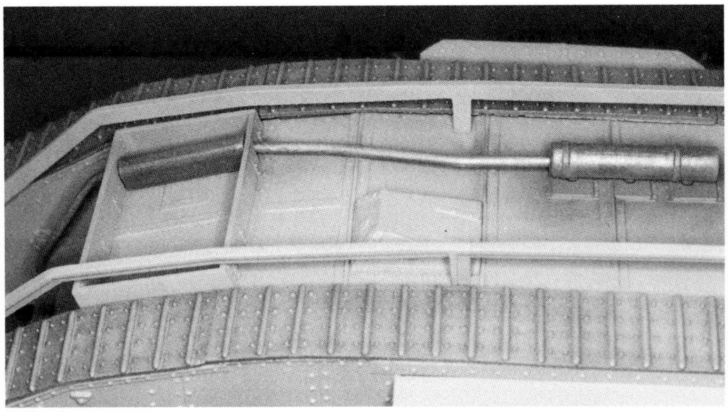

Coat the exhaust pipes and mufflers on this British female tank with Metalizer burnt iron and then topcoat them with Metalizer burnt metal. Then touch the parts lightly with 0000 steel wool to remove the surface color and allow the darker burnt iron color to stand out. Also note the slight surface discoloration around the exhaust collector. You can achieve this look with pastels.

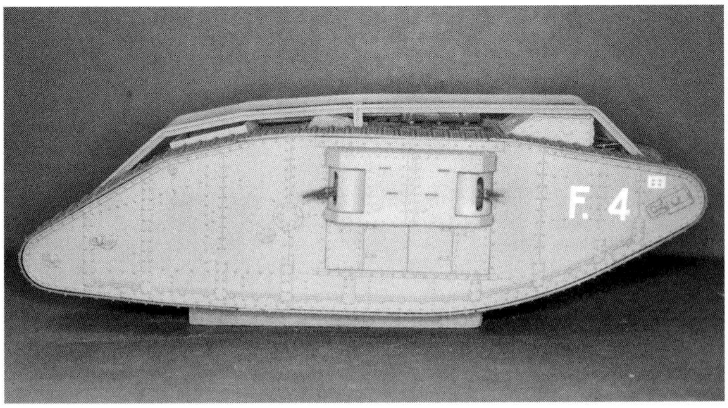

Here's one last example of the subtle weathering effect that dilute washes of water-based mud and dirt colors can have on the overall appearance of a model. Apply very light coats of mud along the vertical rivet lines to help simulate the collection of dirt and mud in the crevices of these areas.

2

Tires, Road Wheels, Sprockets, and Tracks

Giving vinyl tracks a realistic appearance has always been a struggle for armor modelers. Making them appear heavy and wrapped around the sprockets and idler wheels, as well as lying across the road wheels or return rollers, has been one of the great secrets of master armor modelers. Another problem is that the entire tank often appears to be floating on the tracks, especially if they are thick and stiff. Manufacturers have come a long way in the past ten years, and the accuracy and overall appearance of vinyl or flexible tracks have improved greatly. While individual track links can really enhance the appearance and accuracy of your model, they also come with an increased cost—time and effort. Most individual track links are not designed to snap together, so each link has to be individually set and positioned.

Tires. Vinyl tires have also been a problem. Getting rid of the casting seam line can be difficult because the seam line, like the tire, flexes. There is an easy solution: place the tires in your freezer for 24 hours. While still slightly flexible, the seam line will then respond well to a medium- to fine-grit sanding stick. Achieving a fine demarcation line between the rubber and the metal rim on plastic tires has also been a problem for many armor modelers. Read on for an easy solution. It's the same one aircraft modelers use on landing gear tires.

Kits come with two basic types of tires. They are either two-piece injection-molded tires with the hubs already part of the tire, or vinyl tires that have wheel hub inserts. Injection-molded tires are assembled just like any other two-part assembly, but you need to pay particular attention to the alignment pins—I have found that they can sometimes be off enough to offset the tire fit. I usually remove the pins, flatten the gluing surfaces, tape the parts together, and then apply a thin bead of super glue with a thin wire applicator along the seam line. Use very small amounts of glue because you do not want it to seep down into the tread area. Careful scraping and sanding results in a perfect tire.

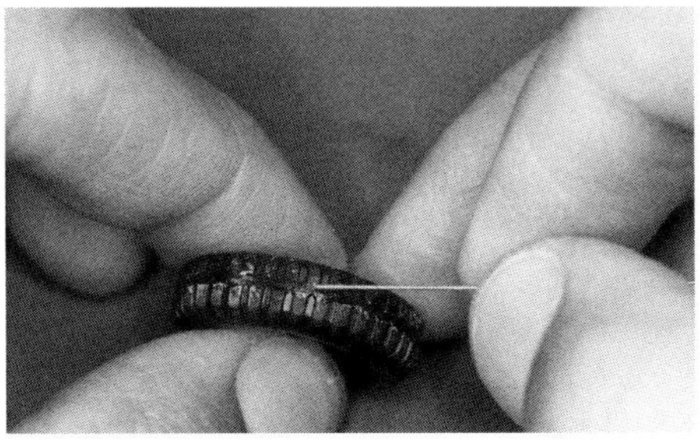

Align the tire halves and then apply a bead of super glue along the seam with a thin wire applicator. Use small amounts—you do not want the super glue seeping down into the tread area of the tire.

I paint the hub and tire whatever color the hub will be, then mask the hub and paint the tire. To give the tires a slightly worn and used appearance and to help them stand out, mix a little flat white with the flat black so the resulting color is a dark charcoal gray.

The first step in working with vinyl tires is to remove the seam along the center of the tire. As I noted earlier, the secret to removing the seam line is to freeze the tire first. While the vinyl will not become completely solid, the seam line will become stiff enough to respond to sanding. Use a fine-grit sanding stick to remove the seam, but be careful not to damage the tread detail. Don't worry about the fine scraped appearance of the vinyl surface along the tire's outside diameter—this will disappear when you paint it. Next, test-fit the hubs and then paint them the required color. Insert the painted hub in place on one side, place some glue on the interior area of the hub, and then position the other hub half. When you are done, give the assembled tire a coat of Testors Dullcote to remove the shine of the vinyl and to hide the surface scratches from the sanding stick. The Dullcote will also lighten the overall color of the tire.

Road wheels. When removing road wheels and return rollers from their sprues, be sure to leave some excess plastic on the part. Trim the excess on a hard surface if it is on the edge of the wheel, and slice it off if the excess is in the middle. The point here is not to damage the wheel or distort its shape. There are usually mold lines along the rim of the wheels and rollers, so be sure to scrape them off carefully using a number 11 X-acto blade. After you have finished scraping, carefully smooth the scraped areas with fine-grit sandpaper. Once you have cleaned up all the wheels and rollers, test-fit them into place along with their axles, suspension arms, and hubs. Next, prime the road wheels and return rollers, check for any defects or missed mold lines, and then reprime any areas that need work.

Tracks. Now you are ready for the part that causes many modelers to get frustrated with anything that has tracks. How do you paint the rubber on the rims of the road wheels and rollers and achieve a sharp demarcation line? The secret is that you do not paint it unless there is a high enough lip between the metal and the rubber so that you can get a small strip of masking tape to stick to it. While a few road wheel designs may have this high lip, the return rollers do not. On the rare occasions when road wheels have this high lip, you can use an airbrush to paint them. In the vast majority of cases, however, the lip between the metal and the rubber is just a raised line and masking tape will not work.

The simple solution is to use a set of disposable drawing pens. They are inexpensive, and you can find them in office supply stores. The ink is permanent and sticks very well to flat paint. Tip sizes can range from .1mm to .5mm. They are easy to use, and the fine tips of these pens will follow the edge of the smallest raised line. I use these pens to set the demarcation line between the metal rim and the rubber tire. To fill in the area along the edge as well as the outside diameter around the road wheel or return roller, I use a felt-tip permanent marking pen. The tips of these pens are pointed, so I simply cut the tip so that it looks like a small flat brush. When using these pens, work in one direction and do not go back over a freshly inked area while the ink is wet. Let the ink dry, which usually takes only a few minutes, and then give the area another coat. The ink will have a slight sheen to it, and the color of the ink will not appear to be uniform. Don't worry—the sheen and the uneven appearance of the ink will disappear once you apply a coat of Testors Dullcote.

When removing sprockets and idler wheels, don't break any of the teeth on the sprockets or distort the rims of the idler wheels. Typically, the sprocket is attached to the trees right at the tip of one or more of the sprocket's teeth. It is best to snip the trees around these parts first, then snip away the remaining tree plastic, leaving some excess at the attachment point, and then carefully remove the excess with the tip of a number 11 X-acto blade. When attaching the two halves of the sprocket wheel, test-fit the track around it to be sure that the teeth on both sides line up into the holes in the track.

Occasionally you will find a kit in which the locating pins on the sprocket part halves are slightly misaligned. In this case, remove the locating pins, adjust the parts, and then glue them together. Most tracked kits have vinyl inserts that go between the road wheel, sprocket, and idler part halves. These vinyl parts are designed to allow the wheels to turn on their axles so you can motorize the kit. Sometimes you can disregard the vinyl inserts, and sometimes you'll have to add them, since the holes for the axles will have some play in them without the inserts. In either case, I glue all the road wheels, sprockets, idlers, and return rollers in place.

Once the glue has dried, scrape the glue and the seamline flat. Light, gentle scraping usually does the job, but be careful not to mar the tread detail.

Vinyl tracks. Tracked vehicle kits like tanks, self-propelled artillery, and half tracks have come with vinyl tracks for many years, and modelers have struggled with how to make them appear to be wrapped around the sprockets and idler wheels and resting or sagging on the return rollers.

There are three secrets to working with vinyl tracks. First, you need to add some weight to the inside of the model so the road wheels rest on the tracks instead of appearing to float on them. Large flattened lead fishing weights are great for this purpose and can be secured with super glue.

Second, test-fit the track by temporarily attaching all the road wheels, return rollers, sprocket, idler wheels, axles, and suspension arms in place to see how much slack or tightness there is on the track. Some kits allow either the idler or sprocket wheels to be adjusted to compensate for this, but in the vast majority of cases you will have to modify their location if you need to adjust the track tension. Fortunately, since most kits are well engineered, this does not happen very often, but once in a while you will come across one that needs adjustment. Be sure not to make the track too tight, or you won't be able to give it the appearance of lying on the road wheels or return rollers. To test the track, you'll have to assemble it permanently. I use either a hot knife or a small-tipped soldering iron to melt and mash down the vinyl pins.

The third and final secret is to select attachment locations on either the return rollers or the road wheels where the track will be tied to these wheels. Mark these locations. Since you will pull the track down onto the road wheels or rollers with clear nylon thread, you'll either drill small holes into the track or use the existing idler sprocket holes as attachment points.

Paint the tracks before you attach them in place, and tie them down. Don't forget to paint the rubber pads if the track has them. Testors Metalizer nonbuffing paints stick very well to vinyl, provided you wash the track to remove mold release agents first. I like to paint my tracks either Metalizer burnt metal or Metalizer burnt iron with a top coat of Metalizer burnt metal. If you go with the burnt iron undercoat, you can make the burnt iron color come through in areas that you lightly wipe with fine 0000 steel wool. This gives the track a two-tone appearance of metal with hints of rust. If the track has rubber blocks, paint them flat black with a small flat brush. The flat paint will stick well to the Metalizer colors if you use the non-buffing Metalizer paints.

Individual track links. These are supposed to be the answer to the sometimes crude appearance of vinyl tracks. Although using nylon thread to pull the tracks down fixes the problem of sagging, individual track links do have an improved appearance. However, it takes a lot of time to clean up and assemble individual track links. You have to be careful how you proceed with the assembly, or you will end up with some strange-looking tracks.

I always start around the sprocket and add one link at a time around it, carefully gluing individual links with small amounts of super glue applied with a thin wire applicator. I am very careful not to glue the links to the sprockets. Once I get the links around the outer edge of the sprocket, I turn my attention to the long flat sections on the road wheels. I carefully tape lengths of tracks together, lay them along the road wheels, and then connect this length to the links around the sprocket. You will need to add a few links between the sprocket's links and the road wheel length, since these are the links that will have a slightly curved appearance. I then repeat the process on the idler wheel and the rear flat run of road wheels. Here again you will need to add a few links to connect the track length that is wrapped around the idler wheel and the rear run of road wheel tracks.

Now you are ready for the track run along the return rollers. You can position another run of flat straight tracks along the return rollers that connect with the tracks wrapped around the drive sprocket and the idler wheel, or you can add a few links if you like sagged tracks.

A word of caution here. Except for just a handful of tracked vehicles that have sagged tracks whether they are running or stopped, almost all tracked vehicles have fairly tight tracks to prevent the tracks from being thrown. Once you have completed assembling and gluing the lengths of track, you should have three or four subassembled lengths of individual track links. Use the same color Metalizer paints and flat black for the rubber pads as you would on vinyl tracks. Finally, position the lengths of assembled and painted track onto the sprocket, road wheels, idler, and rollers and then carefully glue them together. I usually start at the sprocket and work towards the idler wheel.

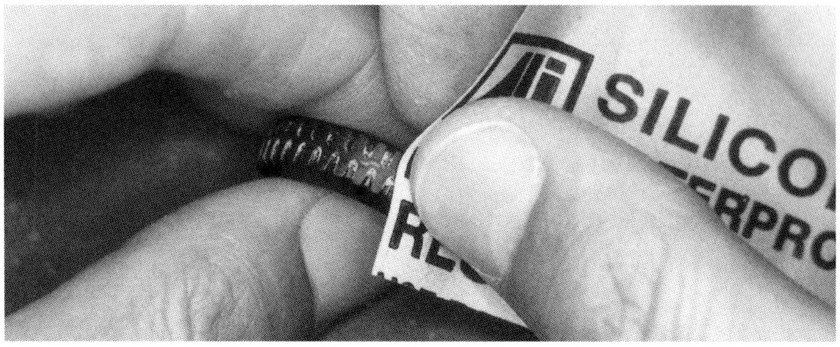

Smooth out the plastic and the glue using fine-grit sandpaper and then wet-sand the plastic to remove any surface scratches.

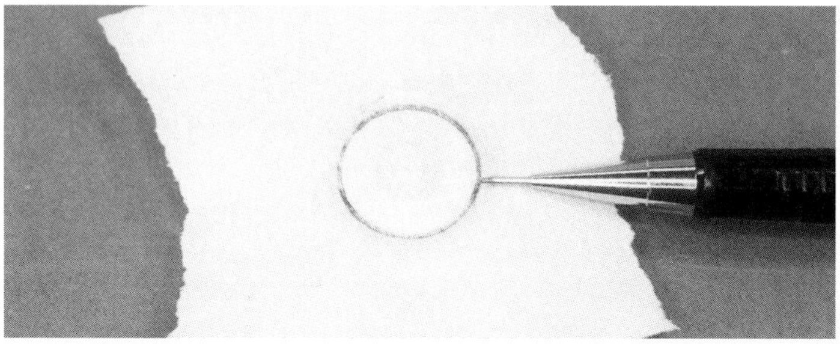

To paint the tires, airbrush the hub the necessary color and then mask it. Cover the hub with a length of masking tape and then draw the raised demarcation line between the steel rim and the tire onto the masking tape. Painter's masking tape stretches a little, so the tape will lie down on the raised area of the hub as you draw.

Next cut out the masking tape around the rim with the tip of a number 11 X-acto blade. Go slow and follow the line carefully.

Paint the tire and then remove the masking tape. If you were careful when you cut out the tape, you should have a perfect demarcation line. You can always touch up the edges of the tire where it meets the steel rim with a detail brush if you have some bleeding.

Vinyl tires have mold lines that have to be removed, but because the seam line is so flexible it will not respond to sanding. The trick is to freeze the tires and then sand them. While the vinyl will not become completely solid, the mold line will respond to sanding.

To assemble vinyl tires, paint the hubs and then assemble them onto the tire. To lighten the color of the vinyl and to dull its shine, give the completed tires a coat of Testors Dullcote.

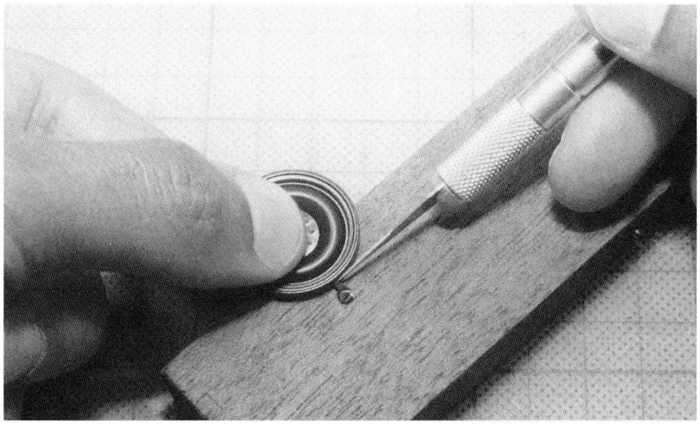

Always cut parts from their trees leaving a little excess plastic on the part. If the excess is located at the edge of the part, trim it off on an elevated hard surface. Be careful not to distort the shape of the parts.

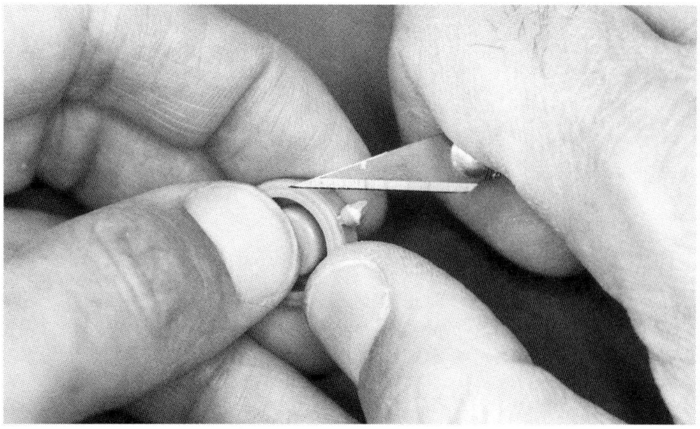

If parts have stem connections in the center, you'll have to slice them off without the benefit of a hard surface. If you lay these parts down and slice off the excess plastic, you stand a good chance of also removing some plastic from the part, resulting in a hole that needs filling.

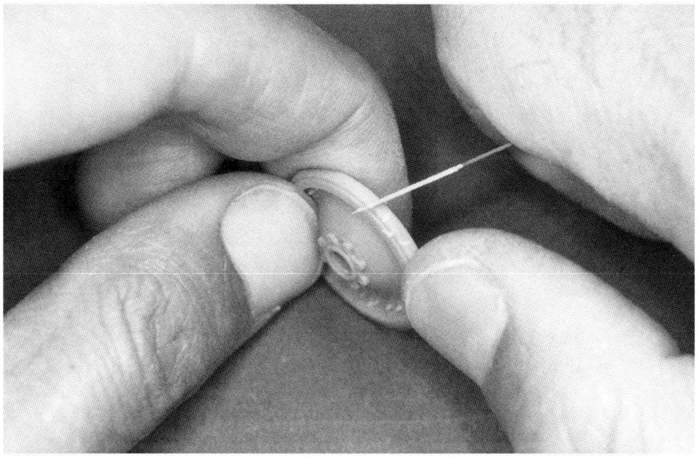

Once you have removed the excess plastic, scrape the surface smooth. Don't forget to remove the mold line around each road wheel as well.

While some mold lines are subtle, other will stand out if you do not remove them. On thin disk-type road wheels, just scraping the line flat will do, but on wide wheels like the ones you find on a Sherman tank, you will have to carefully scrape off the mold line and then completely sand the surface smooth.

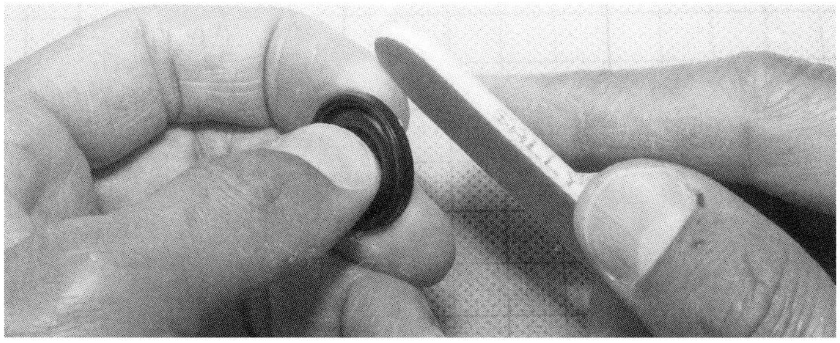

To remove any excess plastic that the knife blade might have missed, gently hit the excess plastic with a sanding stick.

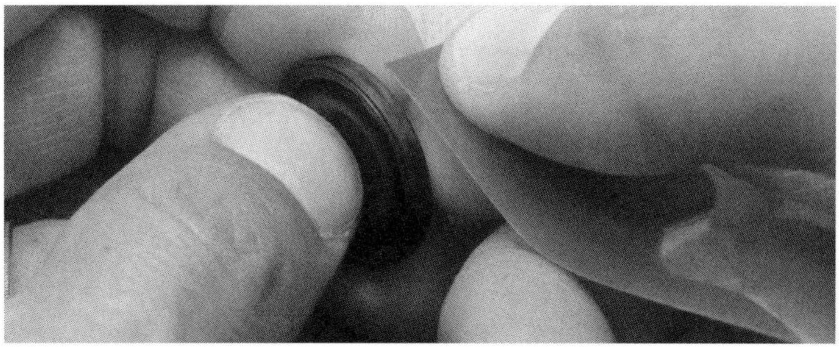

To smooth the plastic and polish the surface, use fine-grit sandpaper and wet-sand. To ensure that you do not distort the shape of the wheels, wrap the sandpaper around the wheel as you sand.

Once you have completed cleaning up the road wheels, idlers, sprockets, and return rollers, check the fit of all the parts, including the suspension arms and axles.

Also check the fit of the axles and suspension arms in the hull. Sometimes these parts do not always fit snugly into their locations. The time to fix these problems is long before you have painted all the components and are ready to assemble them.

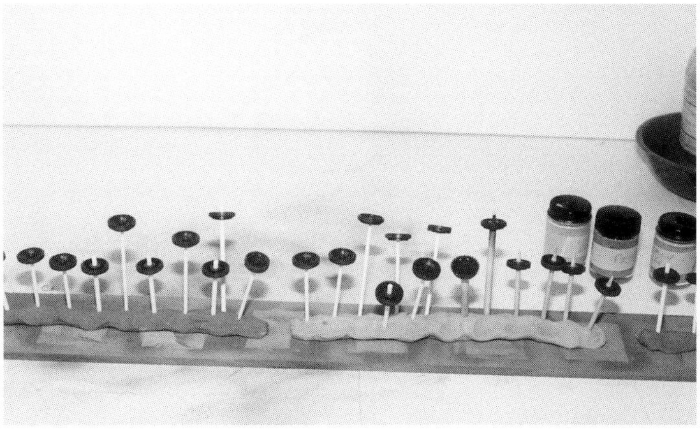

To paint all the wheels, roller sprockets, and idlers, I place each one on a plastic or wood dowel so that I can paint both sides at once. This saves a lot of time.

The dowels also allow you to rotate the wheels as you airbrush. This helps ensure that all the raised surfaces are painted.

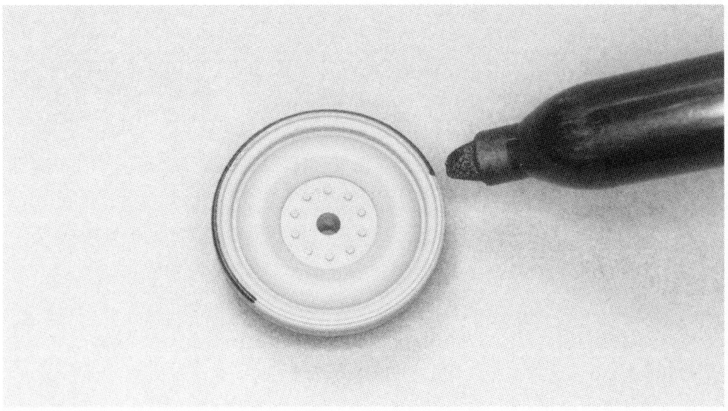

The secret to painting the rubber tread on road wheels and return rollers is not to paint them, but ink them. Indelible black ink felt-tip and needlepoint-tip pens are great for painting the rubber tread. Needle-tip drawing pens come with .1mm to .5mm tips. These fine tips will follow the shallowest raised line on any road wheel.

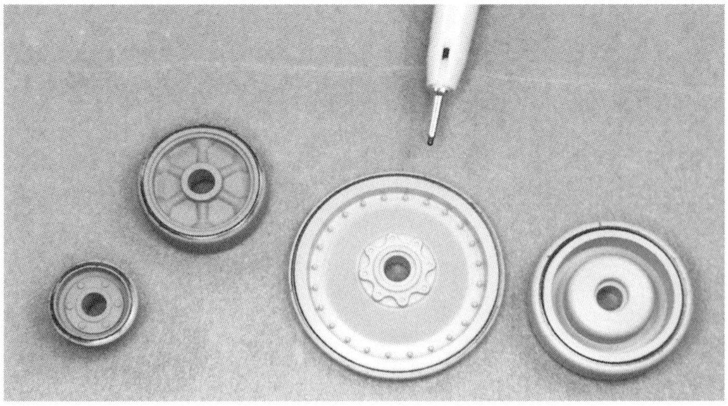

Use a fine-tip disposable drawing pen to set the demarcation line between the rubber and the steel. Then use progressively thicker pen points to fill in the remainder of the tire.

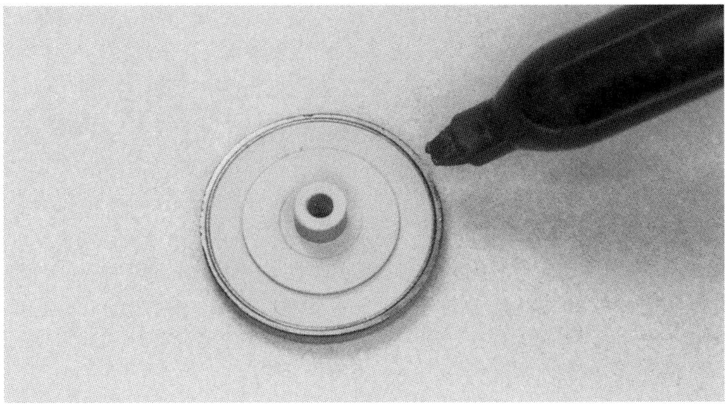

Use a thick-tip felt drawing pen with the tip cut flat to ink in the area around the edge of the rubber.

Both edges are complete on these road wheels, and they are ready to have the rims inked. Note the fine demarcation between rubber and steel. Using drawing pens is quick and easy, and there is no mess.

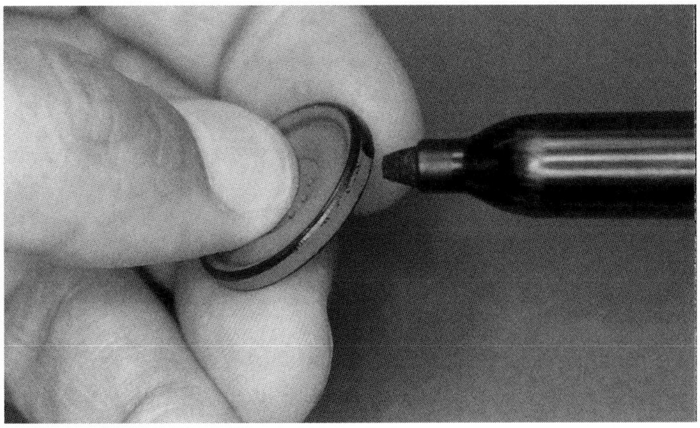

Larger felt-tip pens with flattened tips are great for inking in the rims of the road wheels.

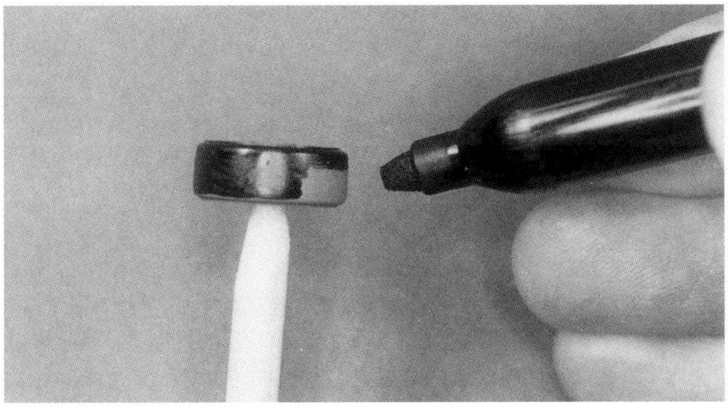

For small road wheels and return rollers, put the part on a dowel and rotate the part to speed up the inking process.

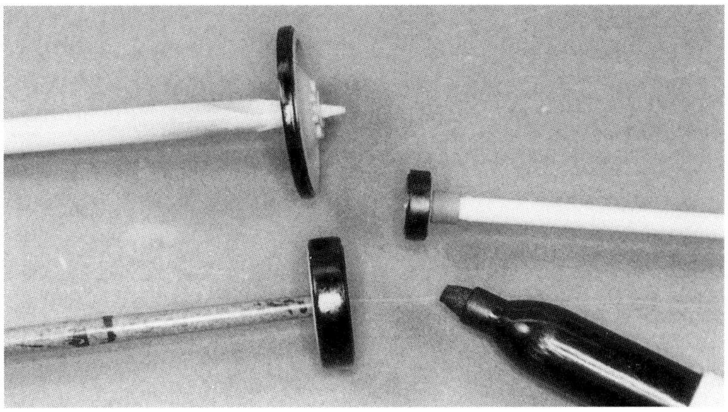

These road wheels and rollers have been completed. The inked surfaces will have a tendency to appear streaky and shiny. To even out the surface appearance and remove the shine, give the wheels a coat of Testors Dullcote.

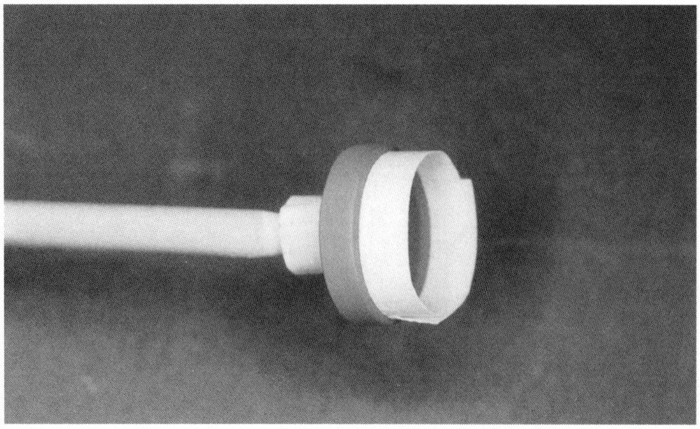

Some road wheels have a high lip separating the steel rim from the rubber, which will enable you to use masking tape. Most road wheels and return wheels have very shallow raised lines, but occasionally you do get road wheels with these high lips.

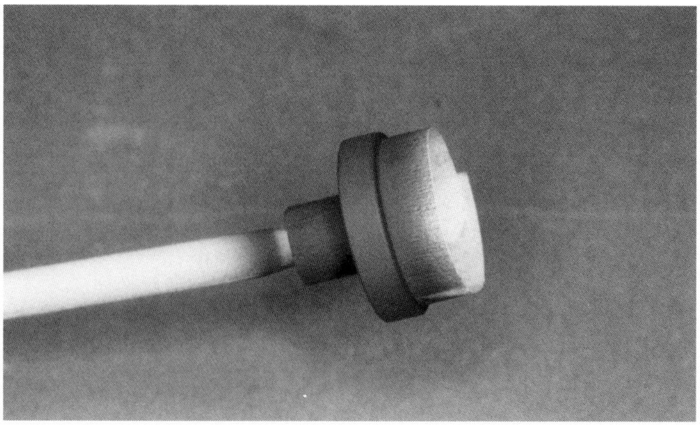

The next step is to paint the road wheel flat black. Add a few drops of flat white to the flat black to lighten the color of the rubber so that it will appear worn.

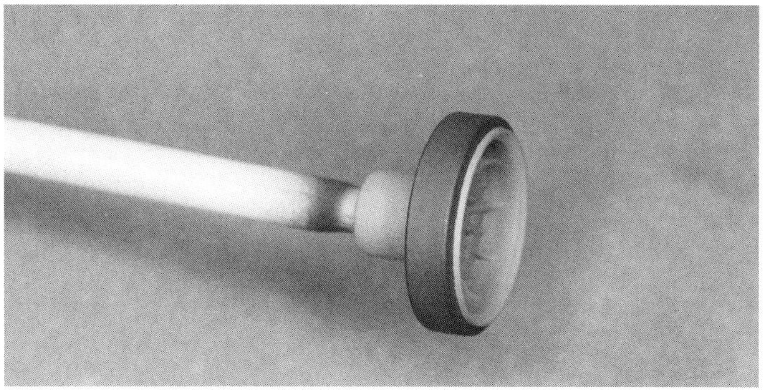

Here is the completed road wheel. Note the fine demarcation line that good masking can achieve.

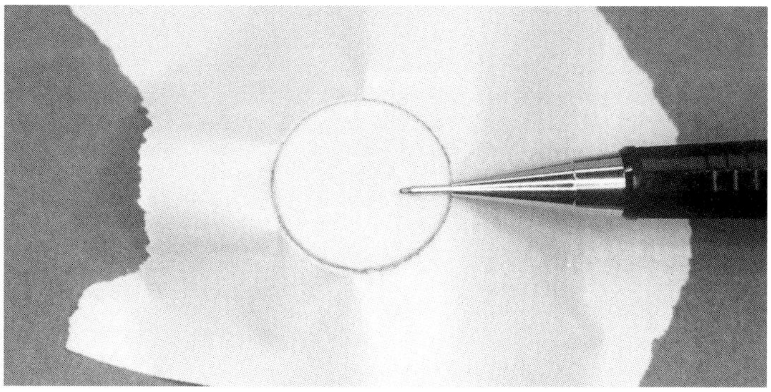

Sometimes you can also use the same previously illustrated masking technique for road wheels that you use for regular wheels.

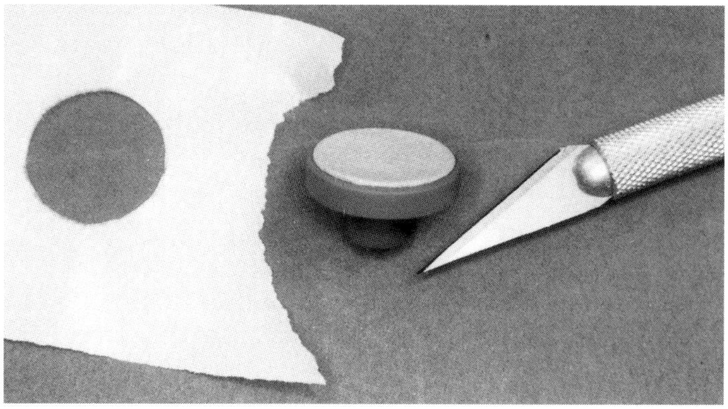

Next, carefully cut out the excess masking tape and then proceed to airbrush the parts.

Occasionally, manufacturers get resourceful in trying to help modelers deal with long-standing problems. Several companies have experimented with separate vinyl inserts for road wheels. Be careful when removing the excess vinyl from these delicate parts. If you gouge the vinyl, position the completed road wheel so the gouged area touches the tread, and it won't be so noticeable.

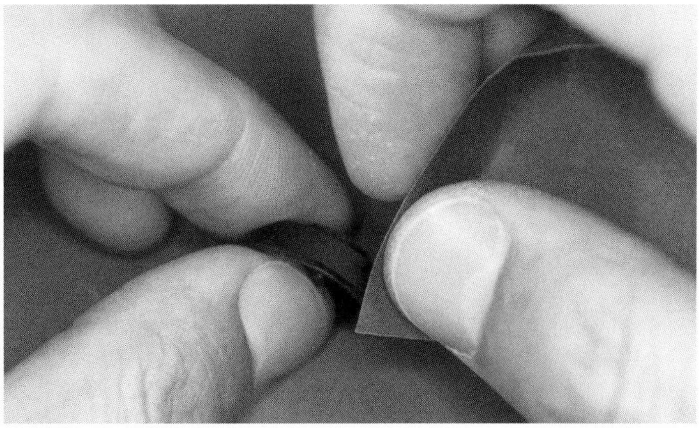

To sand the vinyl smooth, position the vinyl onto the road wheel before sanding so the vinyl won't flex, and be sure to freeze the part before sanding.

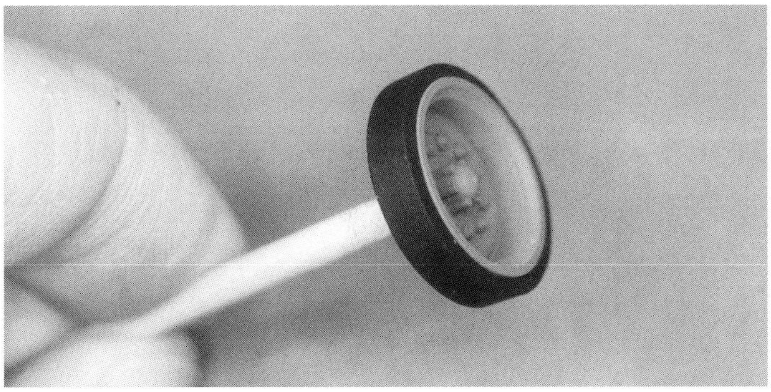

Attach the vinyl insert to the road wheel and give the completed part an overcoat of Testors Dullcote.

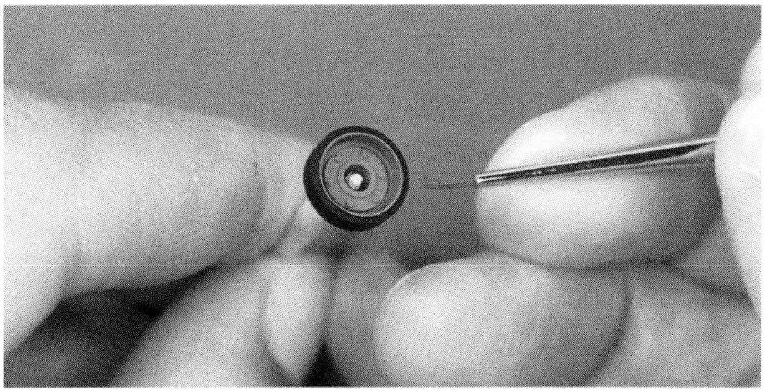

When using fine-tip drawing pens, you will occasionally slip. Simply touch up these areas with a fine detail brush. Paint will cover the unintended inked areas very well.

These completed road wheels and return rollers now are ready for installation. Each one of these road wheels has a raised line of a different height between the rubber and the steel, yet all the demarcation lines are clean, thanks to the fine-tip drawing pens.

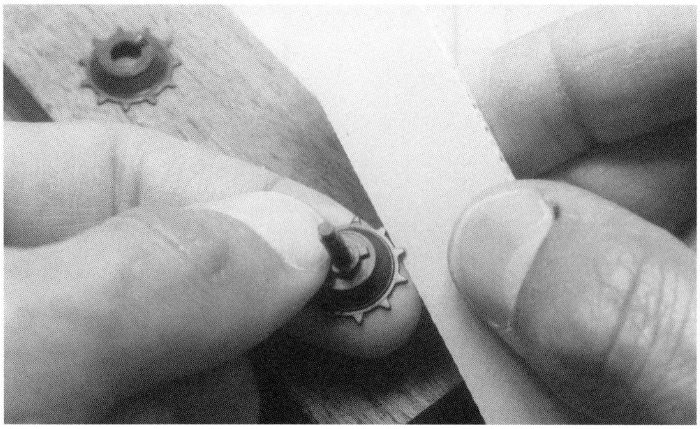

Sprockets are usually attached to their trees at the teeth, so you'll have to be very careful when removing these parts. Careful sanding with a sanding stick can easily restore the shape of the teeth.

Glue together the multiple pieces of these idlers and sprockets, and fill the seams with super glue.

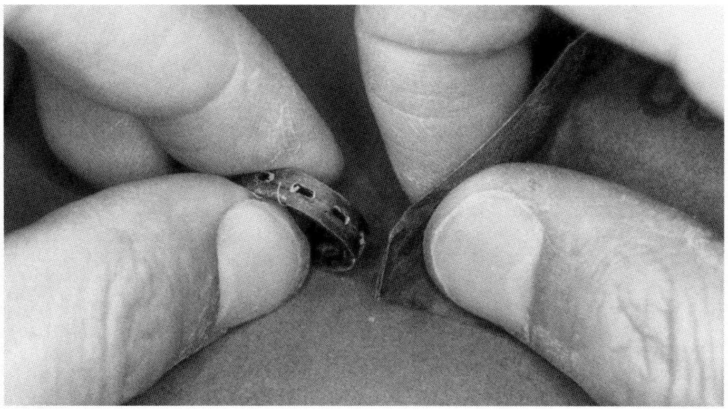

Careful scraping and sanding will help ensure that you don't distort the parts.

The completed sprockets and idlers are ready for painting. The seam lines are barely noticeable. This is one of the great advantages of using super glue. When it dries, it can be sanded and scraped like plastic and it blends in perfectly.

For vinyl tracks that are attached to trees, cut the track plus some excess and then carefully trim away the excess on a hard surface, just as if the vinyl track were a plastic part. You only get one shot at trimming vinyl, because it is almost impossible to repair, so do it right the first time.

To help ensure that the vinyl pegs fit securely and firmly into the corresponding holes on the tracks, use a small-diameter dowel to push the track halves together. Even a pen will work.

Instructions usually call for the modeler to burnish the ends of the vinyl pins with the tip of a hot screwdriver, but a hot knife or soldering iron with a micro tip works best. Where possible, test the hot knife or soldering iron before you use it on the track.

Carefully burnish the tips of the vinyl pins so they flatten out. Be careful how you position the hot knife—do not melt the track!

This track is ready for fitting, painting, and then final assembly.

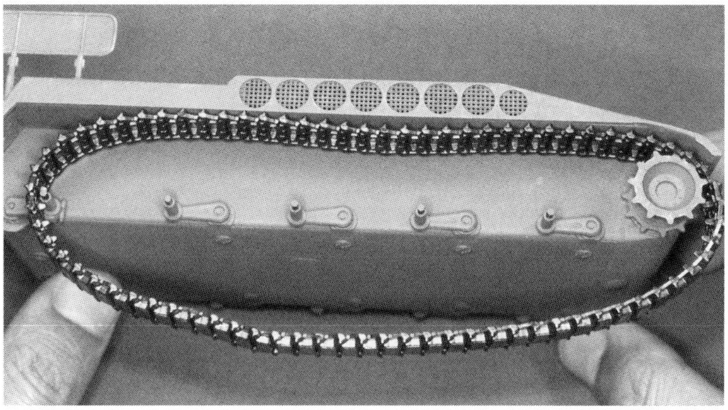

To test-fit a vinyl track, attach the sprocket and return rollers (if the tank has them), and then slip the track onto the sprocket.

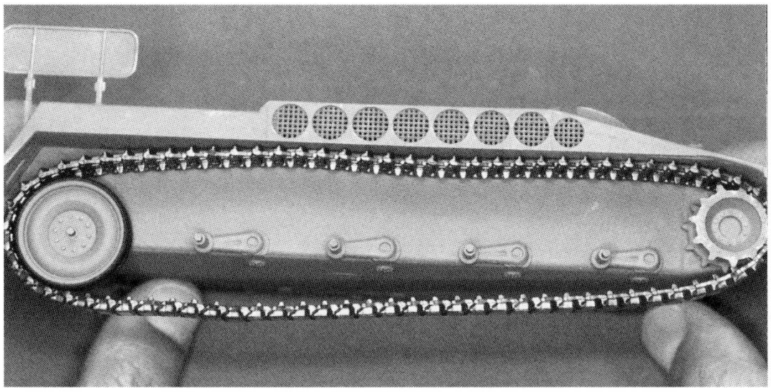

Next, attach the idler wheel by placing the wheel against the inside of the track and then pulling the track taut so that the idler wheel slips onto its axle. This is easier than attaching the idler wheel and then trying to stretch the track over the idler wheel.

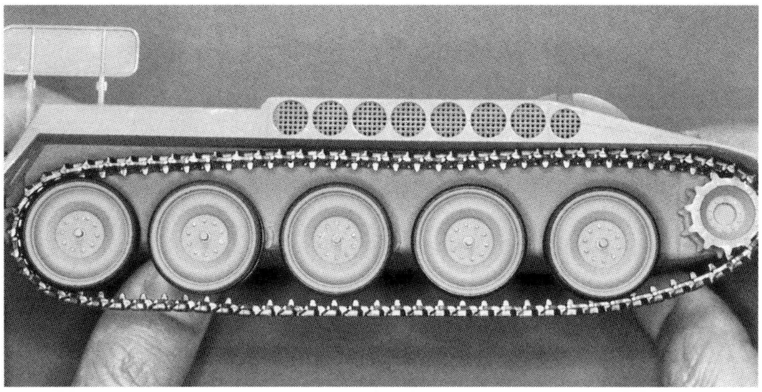

Next, attach the idler arms and the road wheels. The track on this model has plenty of slack. It will be easy to use nylon thread to pull the track down onto the tops of the road wheels.

Testing the track on this model identified the fact that the suspension arm on the idler wheel was bent inward because of the way the track sat on the wheel. This skewed idler wheel was very noticeable and had to be corrected.

The simple solution was to add small pieces of Evergreen strip stock to the inside areas of the suspension arms.

Glue the idler wheel suspension arms into place, and you'll just have to paint the plastic strips.

Testors Metalizer burnt iron and metal are great paint colors for tracks. To paint the rubber pads on tracks use a small, flat brush.

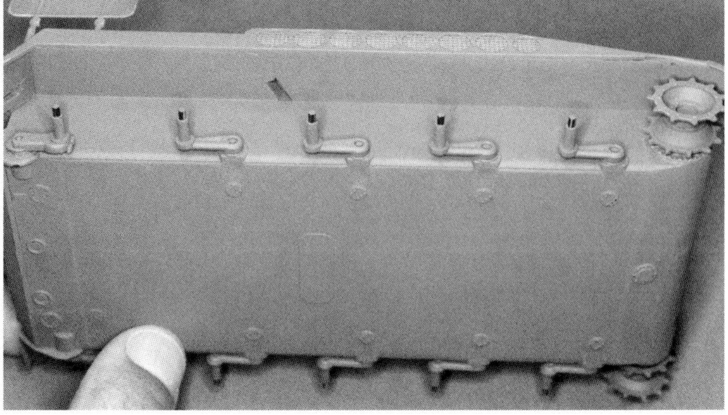

Once the tracks are painted, the next step is to permanently install the sprockets, suspension arms, and return rollers.

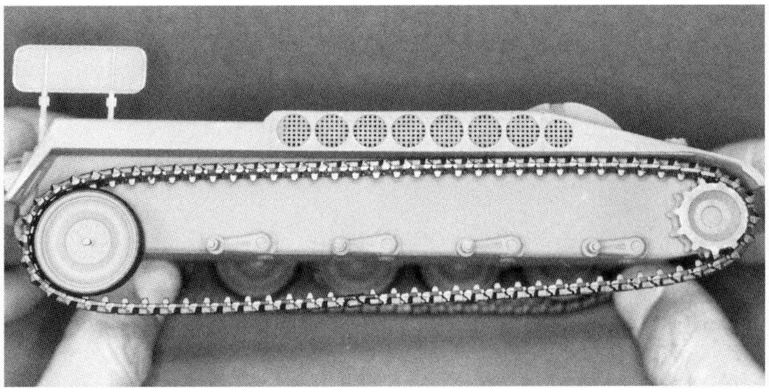

Next, attach the track to the sprocket, insert the idler wheel on the back half of the track, and then stretch the track out until the idler wheel fits into its location.

Next, drill small holes in the vinyl tracks to accept clear nylon sewing thread. Each location should have two holes on opposite ends of the outer edges of the track so you can run the nylon thread through both ends, giving the thread a secure attachment point. The thread locations must correspond to the locations of the road wheels.

Install the road wheels and wrap the thread around the area between the inner and outer wheels. Pull the thread tight until the track lies down on the road wheel.

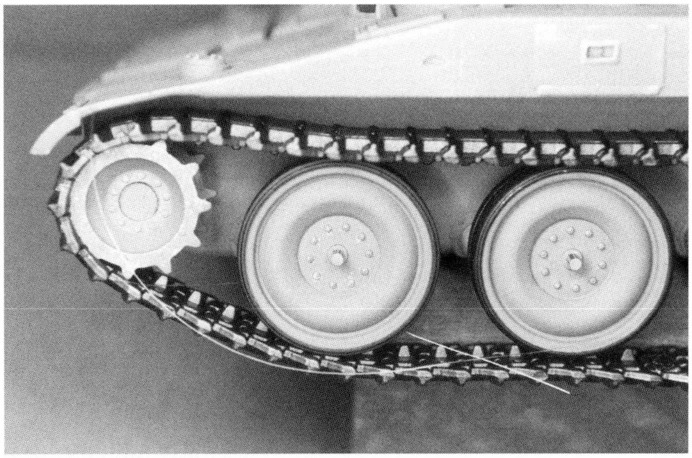

Knot the nylon thread and cut it. The track will lie on top of the road wheel.

Tie the road wheels to the track using a clear nylon thread along the entire length of the track, avoiding a wavy appearance.

Cut the clear nylon thread, and there is no indication that you have tied the upper track to the road wheels.

Although these vinyl tracks are probably slightly thicker than they should be, the overall appearance of the track is better now that the track appears to be lying on the road wheels instead of floating above them.

The drive sprocket on this World War I British female tank had to be trimmed because the sprocket pushed the track too far out. Test-fitting the track around this sprocket helped identify the problem before it was too late. After you lay the track across this area, the cut-up drive sprocket will not be visible.

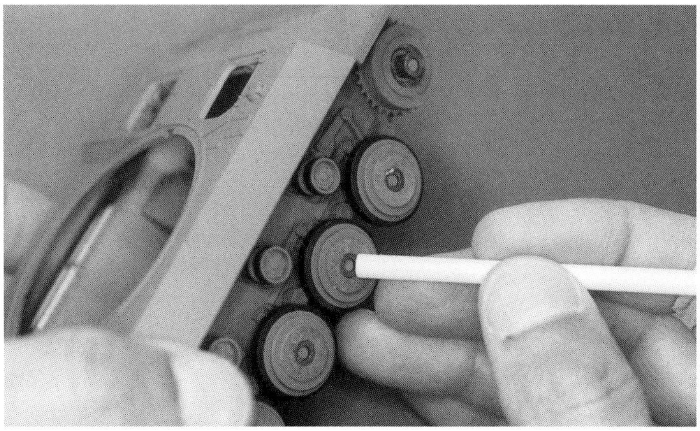

A length of Evergreen plastic tube works well to push vinyl inserts into place on the inside road wheels of this M-18 Hellcat.

Install the drive sprockets, idler wheels, return rollers, and road wheels of the M-18 Hellcat. This tank is now ready for its individual track links.

Cut individual track links directly off their sprues using a sharp number 11 X-acto blade.

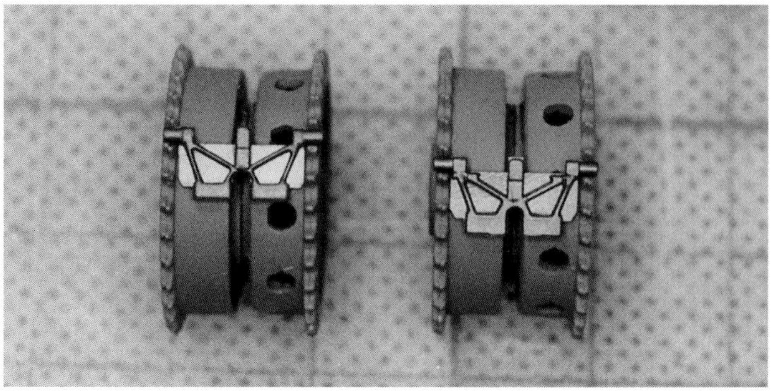

I like to build up individual track links starting around the sprocket. I position the first track link in place and then add additional links, gluing as I progress around the sprocket.

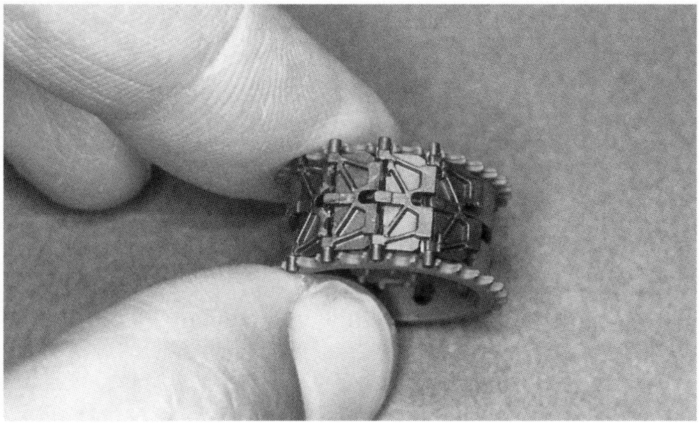

Fit the individual track links around the sprocket and glue the links to each other. Be careful when gluing links around the sprocket or the idler wheel—you do not want to glue the links to these parts at this time.

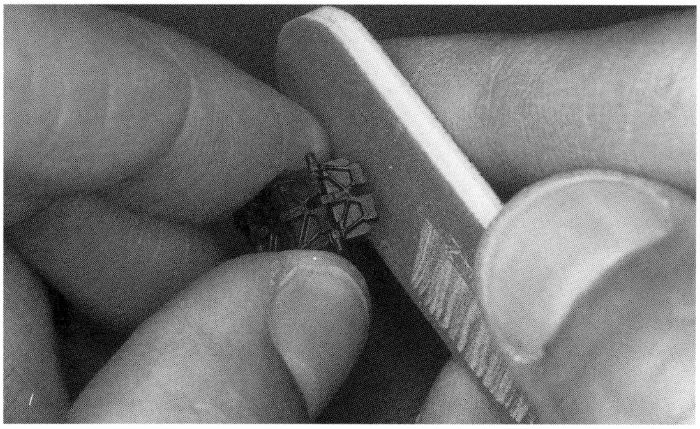

Sometimes you have to sand the edges of individual track links to get them to fit together correctly or tightly. This is most easily accomplished with a sanding stick.

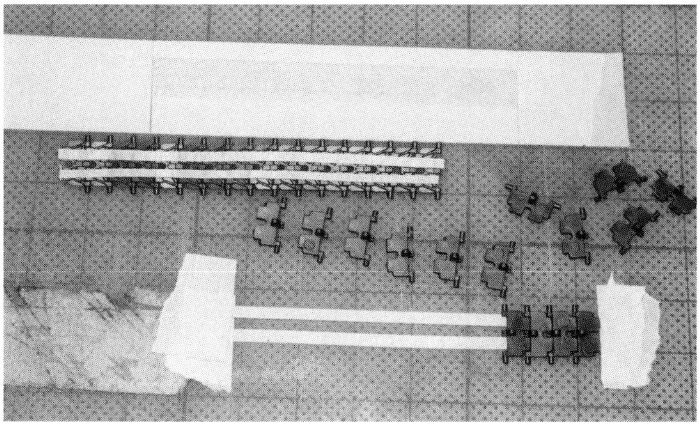

For straight or curved lengths of individual track links, lay out two thin strips of masking tape and then build up the links along the masking tape.

I like to work my way from the sprocket to the road wheels and then to the return rollers and the idlers. Fit a taped length of links from the lower end of the sprocket to the idler wheels. After you position it correctly, you'll glue the individual links to one another.

The forward area of the individual track links is complete. Now it's time to work towards the idler wheel.

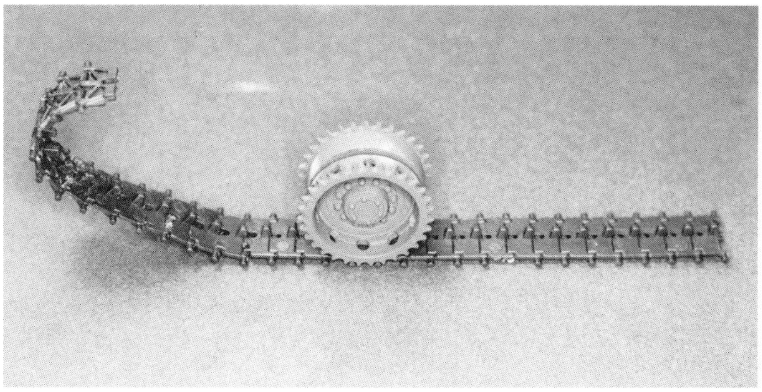

Here is a good example of a length of individual track links that was fitted around the sprocket and the forward road wheels.

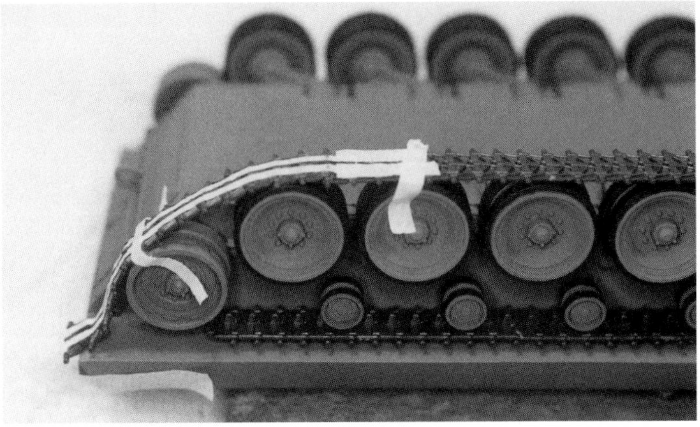

The process of taping individual links together and then fitting them in place around the idler wheel is the same as the one used on the sprocket.

Slow and methodical test-fitting will usually identify problems. While the angle of the individual track links on the upper part of this sprocket is correct, you'll have to add one more link to make the connection between the sprocket and the length of links along the return rollers.

Going slow and test-fitting is important because sometimes you do make mistakes. Here the upper links around the top of the idler wheel have too much of a curve. You'll have to break them off and reattach them at a lower angle.

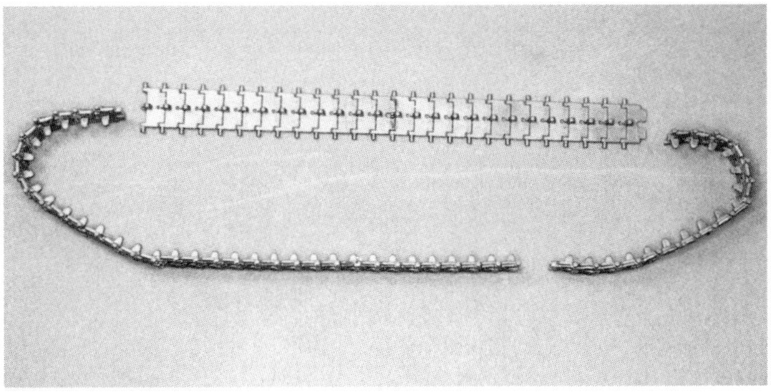

Here is a good example of a set of completed individual track links. The next step is to paint these three sections and then install them on the tank.

Paint the links, position them, and glue them into place. While these tracks look good, there is a slight upward bow between the sprocket and the first return roller. The solution here is to reposition the links by breaking them and reattaching or hiding the problem.

Since this M-18 Hellcat had forward and rear skirts, it was not necessary to make any adjustments to the track. Also note that all the running gear have received their first few subtle coats of mud and dirt wash.

3 Building and Detailing Tanks

Let's turn our attention to assembling the parts on the tank's lower hull. First, deal with any fit problems you've identified during the initial fit check and then move on to attaching parts to the inside of the upper hull. Many of the armor kits manufactured in the Far East have openings in the tank's bottom to accommodate the motorized versions. The easy way to close these areas is to glue a thin sheet of Evergreen sheet stock to the interior, flip the part over, level it, and insert some resin in the opening. Protect the surrounding plastic with masking tape and use a syringe to apply the resin. These plastic pointed applicators are available from Micro Mark. After the resin cures, which takes only a few minutes, sand the surface smooth and apply some super glue around the perimeter of the resin so it will bond to the plastic. If you use super glue, capillary action will pull it down between the resin and the plastic.

Interior detail. Cottage industry manufacturers have been producing multimedia kits of turret, crew, and engine compartments that really add extra detail for modelers who like to display their models with all the hatches open. These resin, white metal, and photoetched kits are highly detailed, but almost all of them require some type of alteration to the kits' interiors before they will fit correctly. Often, their instructions are not clear, so be sure to study all the parts carefully. The actual assembly sequence may vary from what the instructions suggest, so go slow and check the fit of all the major interior components prior to painting and assembly.

Kit manufacturers have also begun to produce kits with either partial or complete interiors. As you build up the interior, check the fit of the parts even if they were supplied by the manufacturer. They sometimes need tweaking. After assembling all the interior parts and completing the fit checks, you are ready to paint, weather, and attach the parts. If you are adding a resin interior, it will be a bit of a trick to add the lead weight that I mentioned in Chapter 1 to help the road wheels sit on the tracks correctly. If it is an engine compartment or the crew compartment, add the weight in the empty area; if it is a full interior, the resin will add some weight, but you'll still have to add some lead. You can hide lead pieces in the space between the bottom of the tank and the floor, under the seats, or in places you will not be able to see once the upper hull is attached. If you are building up a kit interior, glue some pieces of lead weight to the inside of the engine or the transmission before you glue these pieces together.

Painting and weathering. Tank interiors were almost always light in color to help brighten these otherwise dark, damp, smelly places. Anything from flat white to light gray will work. Remember that the lighter the color, the easier it will be for the human eye to see the interior details through the hatches and access covers. To further help pick out the detail, I suggest that you paint the interior parts different shades of the same color or use flat white on some and a flat light gray or a light sea gray on others. Aircraft modelers use the same trick to help pick out the detail when they're painting a cockpit interior of all one color.

Tank interiors are usually dirty and grimy; the transmission and drive shafts have hints of lubricant and oil leaks, and the engine and its compartment are oily and sooty. Use dilute washes of Polly-S water-base weathering paint to simulate oils and lubricants and dark gray and light black pastel pencil dust for the dirt, soot, and grime. If you want to make the oil and lubricant stains more realistic, give these parts a coat of clear gloss first. The dilute water-base paints will then flow into the joints and connection points more easily. If you decide to add the clear gloss, be sure to topcoat it with Testors Dullcote—the pastel pencil dust will not stick well to a gloss surface. Give the complete interior a coat of Dullcote to seal the pastel dust.

One final note here: reference material and photos may suggest that a tank type had a certain interior color. Remember that as the war years dragged on, the choices of paint colors became limited, especially for the Axis powers. (Exterior colors were more consistent.) That's why I suggest flat white to light gray. The interiors might not even have been painted. In some cases, Russian factories assembled tanks and drove them unpainted directly into battle. In the early years of the war, the fighting might have been only a mile or less from the factory—or, at Stalingrad, right outside the factory doors. While the Russians were great at getting the steel to make tanks, painting them the correct interior or exterior colors—or painting them at all—was not always at the top of the priority list.

Exterior assembly and detailing. Once you have finished the interior it's time to attach the upper and lower hull halves. Position them and then add small strips of masking tape to hold them in place and pull the connection points together. Add small amounts of super glue along the seam lines. Then remove the tape and finish gluing. When scraping and sanding the seam areas, be careful not to distort the shape of a seam line that forms a curve or an edge. Sometimes you'll have to add small lengths of Evergreen strip stock to edges when the parts do not fit exactly together. Apply super glue along the entire perimeter of the strip prior to shaping and sanding so you will not have to deal with a seam line.

Once you have completed the assembly of the hull halves, add parts such as storage boxes and bins, closed engine and crew hatches and covers, hooks, grab irons, and lift points. Paint the completed assembly and then attach the sprockets, road wheels, rollers, and idlers. If you are using vinyl tracks, it may be easier to place the track onto the sprockets and idler wheels first and then attach the road wheels and rollers. Other times it is easier to attach all that stuff first and then attach the tracks. If you are attaching individual track links, position the subassemblies and then glue the sections together. I like to work from the sprocket to the idler wheel. Be sure to attach the correct track to the correct side. There may be slight differences in the positions of the drive components on either side of the tank, so be sure the track that was fitted to the left side or the right side goes onto that side.

Next it's time to add detail parts like tow cables, tools, open engine and crew hatches, periscopes, and wood and metal supply boxes. I paint, weather, and drybrush them separately. Once they are complete, I attach them with small amounts of white glue. You can add these details now or wait until after you have completed the turret. The last thing is to attach the mufflers and muffler covers. To improve the appearance of the mufflers, bore them out with small drill bits and a pin vise. I paint the mufflers using Metalizer burnt iron followed by an overcoat of Metalizer burnt metal. I then use 0000 steel wool to bring out some of the burnt iron color. I also add dark gray pastel pencil around the area of the mufflers to simulate the metal and paint discoloration that hot mufflers cause.

Next I add the decals, if there are any. Then, using Testors silver paint, I drybrush all the edges and areas where hands and foot traffic would wear away the paint. If you have a decal like an American star that goes across raised surface detail, be sure to drybrush these surfaces too. Now it's time to spray the lower areas of the tank hull with dilute mixtures of water-base mud and dirt colors. Don't overdo this. Hints of mud and dirt are better than a mud paint coating that changes the color of the base paint. Once the water-base paint is dry, give the entire surface a coat of Testors Dullcote to seal the decals and tone down the drybrushed silver color.

Tank turret assembly is pretty straightforward—usually the parts are either left and right halves or top/sides and the bottom. Tape the turret halves together and check the fit. Also check the fit of the entire assembly in the tank's turret ring. If you have interior parts to install, check their fit and ensure that you can install the breech assembly after the turret halves are glued. This is usually not a problem, but I suggest you check it anyway. If interior parts are attached to the inside walls of the turret, position these and paint the turret inside halves the same color as the hull interior. Use pastel pencil dust applied with a soft flat brush to weather the interior. The breech assembly should get a special treatment of dark gray or black to simulate powder burns. Also drybrush the edges of the breech, especially the slide, with Testors silver. After you have completed weathering the interior, seal it with Testors Dullcote.

When you have finished the interior, position the turret halves with masking tape and run a bead of super glue along the line. Once the glue is dry, remove the masking tape and fill the seam line areas with the super glue. It may take several applications to fill the seam. It's important to ensure that the turret's seams are filled and that the surface of the turret is not marred if it has casting detail. After filling the seam line completely with super glue, carefully and gently scrape the super glue flat using the tip of a number 11 X-acto blade or an X-acto stencil knife. Then use small strips of 400-grit sandpaper to smooth the seam line; finish with number 600 grit. Don't forget to install the gun barrel mantlet if it fits between the turret halves.

After completing the turret halves, test the turret in its location again to ensure that it fits and turns in the turret ring. Next, assemble the gun barrel. Check the fit very carefully and remove any alignment pins if necessary. The barrel halves should fit perfectly because it is a lot easier to fix a seam line along a thin, long round shape than to have to fix the seam as well as correct any distortions that may result from misaligned part halves. Use small strips of masking tape to position the parts and run a thin bead of super glue along the seams. Once the glue is dry, carefully scrape the glue bead line with the tip of a number 11 X-acto blade. To keep the light scraping from flattening the seam line, rotate the barrel a little to either side of the seam line as you scrape. Once you have finished, wrap 400-grit sandpaper around the barrel and rotate it as you sand. I like to use a twisting motion as I pull the barrel through the sandpaper. Finish off with 600-grit paper to polish the surface. When gluing a muzzle brake together, don't let the glue run along the sides or in crevices—this may mar the appearance of the brake. Be extra careful when scraping and sanding these small pieces. It is easy to distort their shape.

Attach the completed barrel to the turret mantlet. If it is a separate part from the turret assembly, install it in place. Install the breech assembly to the back side of the mantlet and be sure the barrel and the breech form a straight line.

Next it is time to attach the detail parts such as lift rings, grab handles, closed hatches, and cage storage frames to the exterior of the turret. Paint the turret and add the decals. Drybrush edges and raised detail to make them stand out. The last assembly step is to add additional details such as periscopes, grenade launchers, machine guns, and antennas. Don't forget to drybrush these parts, too. You can use the same disposable drawing pens to highlight the periscope faces that you used to highlight the rubber rims of the road wheels. To make the machine guns look more realistic, you can bore out the barrels. I make my antennas out of thin brass wire and paint them black.

The front hull seam line on this M-18 Hellcat will need several applications of super glue and careful scraping and sanding. Be sure to sand the sharp angle that defines the upper and lower hull in such a way that you don't skew the sharp line.

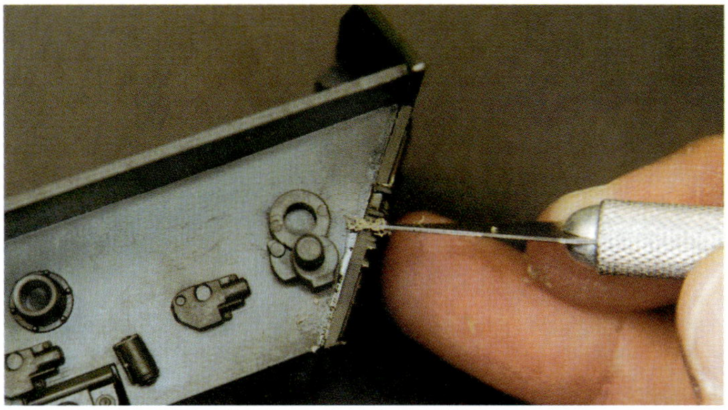

The width of the back panel on this M-18 Hellcat is slightly greater where it attaches to the lower hull. You'll have to reduce it so that it is flush with the side.

After you've finished scraping and sanding, the completed hull panel blends into the side perfectly.

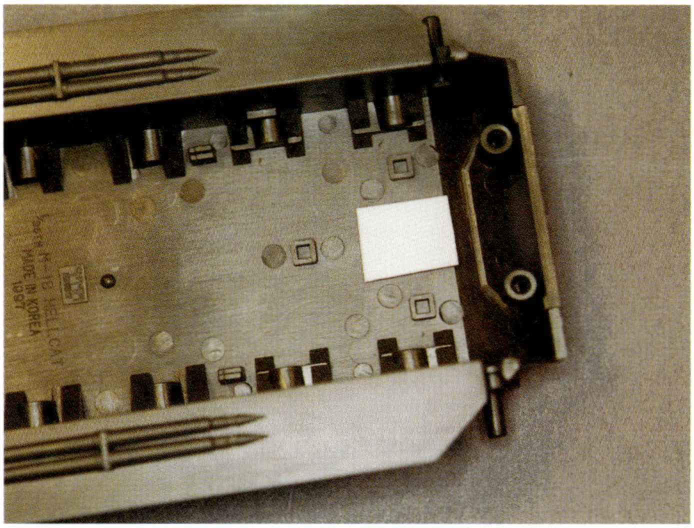

The first step in sealing the motorizing holes that you typically find on tank models is to cover the hole with a piece of Evergreen sheet stock. The next step is to flip the hull over and fill the hole on the hull bottom with resin.

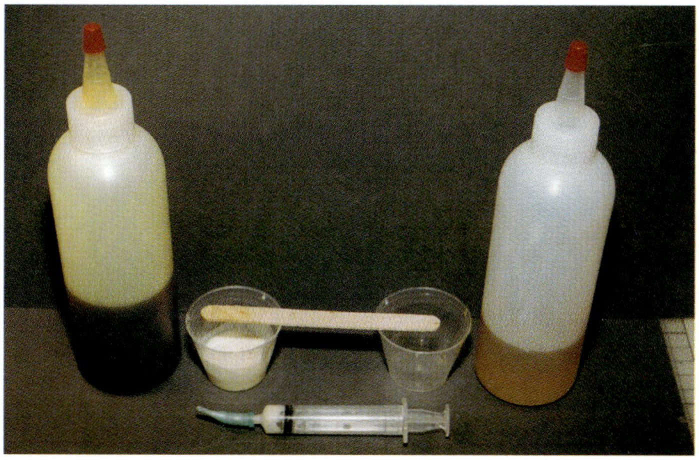

The easy way to mix resin is to use graduated mixing cups of equal volume so that you can measure the resin and hardener. Popsicle sticks are great for stirring the resin mixture. Since you'll pour the resin into a small hole, use a syringe to ensure precise application.

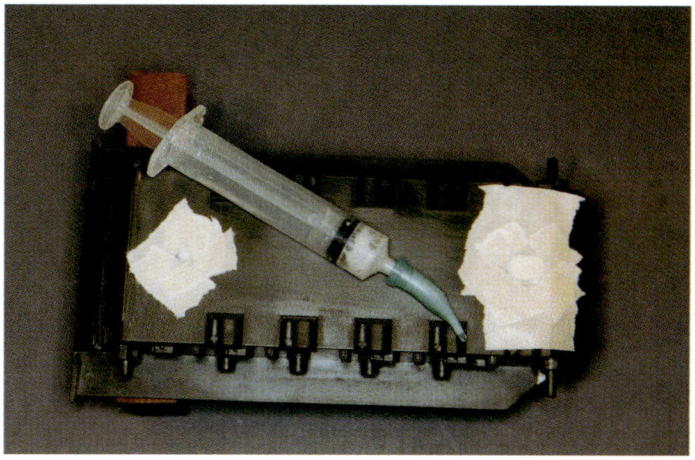

To protect the surface of the plastic, cover the area around the holes with masking tape. Be sure the hull is level so that the resin will completely fill the holes.

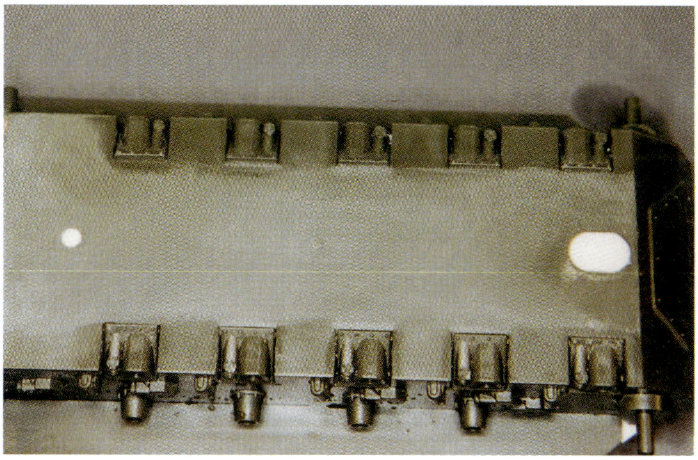

Glue the resin into place by applying a bead line with a thin wire applicator between the resin and the plastic. The capillary action of the super glue will pull the glue down into the seam between the resin and the plastic. Once the glue is dry, sand the resin flush with the hull.

To seal motorizing openings that you sometimes find on the sides of the hull where the drive sprockets are, use strips of Evergreen strip stock. Form-fit the plastic strip into place, apply a bead of super glue along the seam line, and sand smooth.

tips

- Fill holes in the tank hull bottom with resin.
- Use Evergreen plastic form-fitted into place to fill holes on the sides of tank hulls.
- Be sure to test your resin before you use it. Moisture in the air can affect curing.

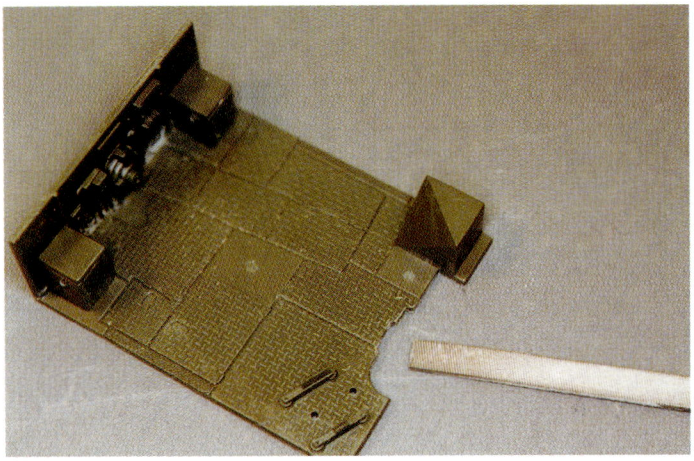

Sometimes you must modify interior parts slightly to get them to fit correctly. Use a file to form-fit interior decking in place, as in this photo.

To form-fit the parts correctly, temporarily install all the main interior components. Modify the decking with a file so that the back end of the transmission plate sits snugly up against the deck.

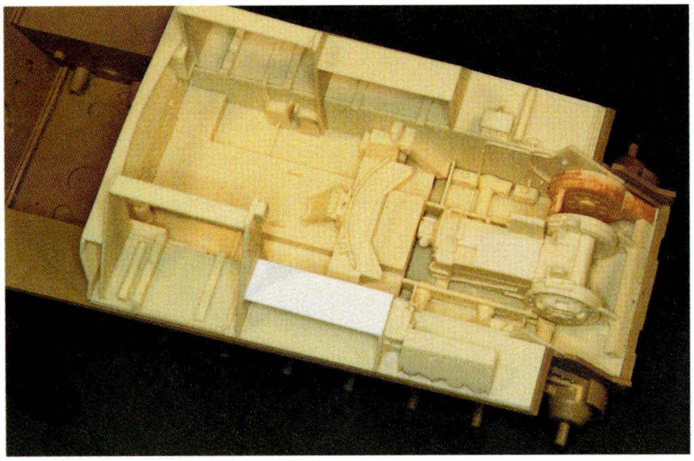

Resin interior super detail sets can add an unprecedented level of realism to any armor model. While these parts are well engineered, they do need form-fitting and tweaking to get everything to fit into place. This Jaguar Jagdpanzer interior detail set is just about complete and is ready for the main components to be painted. Model by Scott Weller

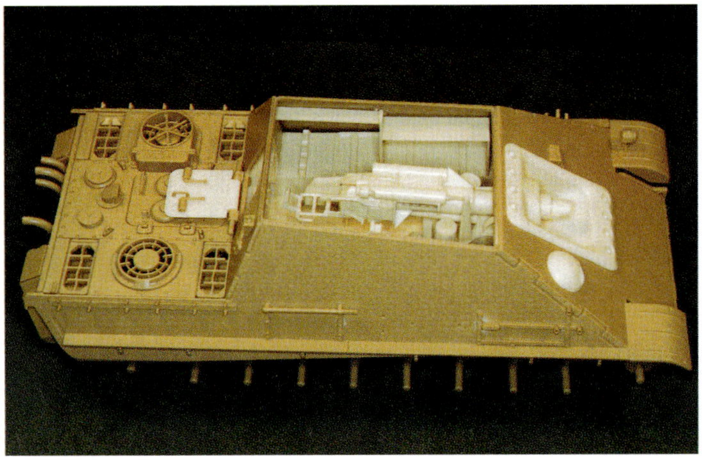

Temporarily attach the upper hull to the Jagdpanzer to ensure that none of the interior parts, including the breech assembly, interferes with its proper positioning. Model by Scott Weller

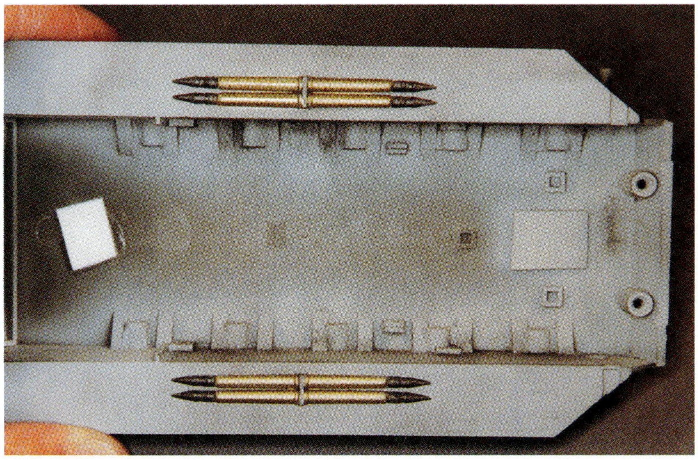

Carefully paint the molded-on projectiles with Testors Metalizer paints and a fine detail brush.

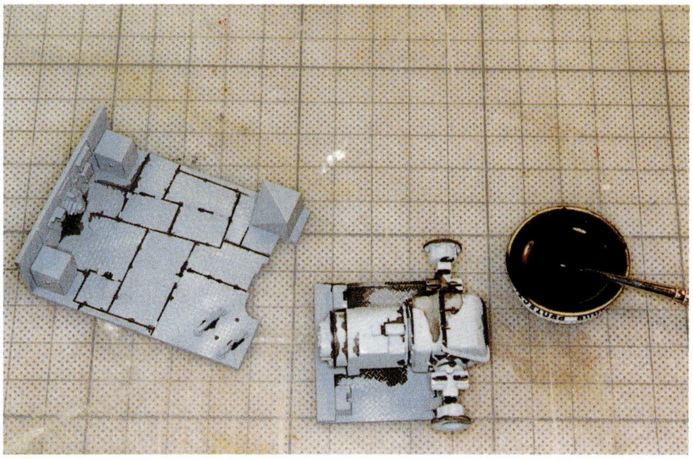

Before permanently attaching the interior parts, finish weathering the subassemblies. Apply a dilute wash of flat black to the corners and crevices of these parts.

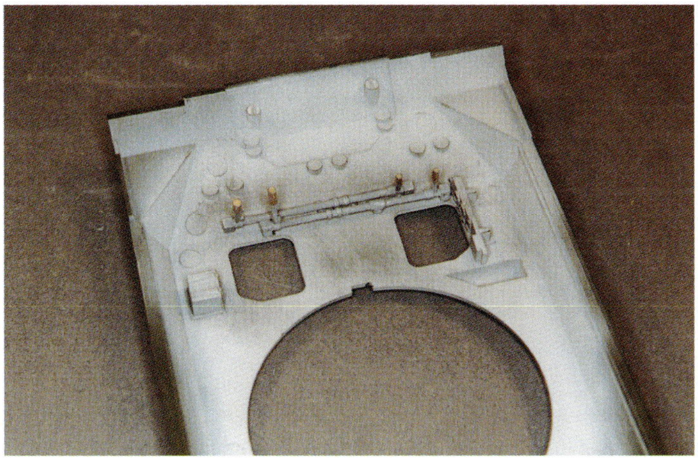

When you're weathering interiors, don't forget to paint and weather the inside upper area of the hull.

Paint and weather the interior parts of the M-18 Hellcat and glue them into place. The next step is to attach the upper hull.

tips

- Resin interior detail parts should be form-fitted into place, especially large interior sides, walls, and decks.
- To add weight to plastic interior parts glue pieces of lead weights to the inside prior to gluing.
- Be sure to weather your interior.

Glue strips of Evergreen sheet stock to the underside of the upper hull of the Russian T34/85 so that the voids on these inside upper areas will not be noticeable. Model by Jim Hudson

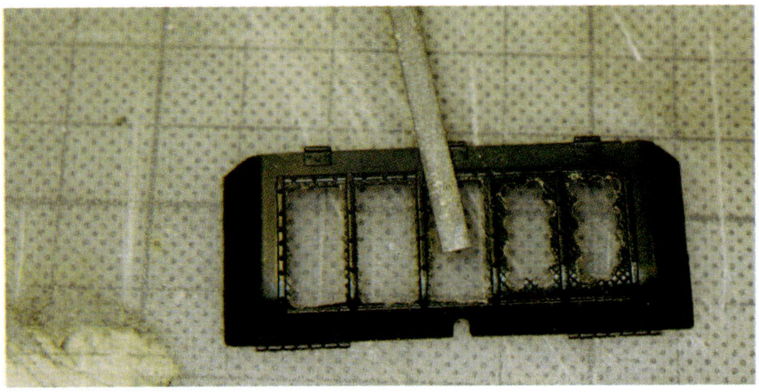

Drill out the molded screen mesh on this engine hatch cover, and then remove the excess plastic with a file.

Clean and smooth the delicate framing with small sandpaper strips cut from sanding sticks.

Install photoetched screening on the completed engine hatch cover of the Russian T34/85. This added detail really enhances the overall appearance of this model. Model by Jim Hudson

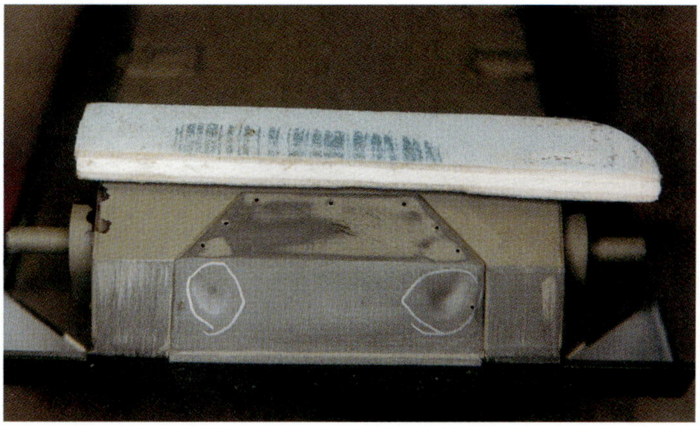

Unfortunately, no one noticed the sink marks on the lower plate of this M-18 Hellcat until after the hull was painted. If that happens to you, cut off the bolt detail on the plating and drill small holes to accept Grandt Line detail bolt heads.

To fill the dimples in the lower plate, apply automotive putty with a flat-tipped X-acto blade.

Wet-sand the auto putty carefully and then polish the surface of the plastic with 600-grit sandpaper.

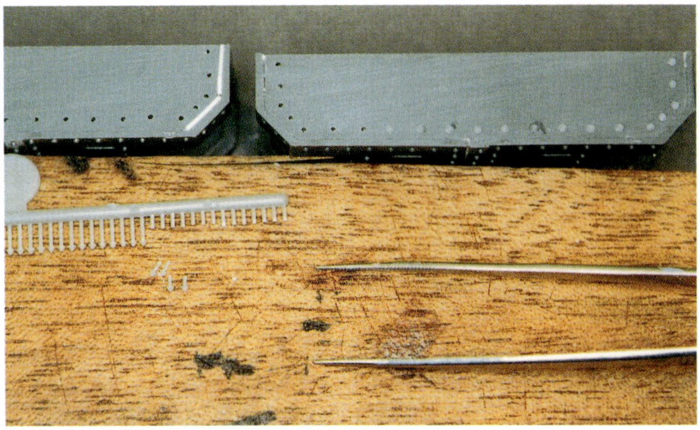

Since these British World War I female tank gun housings do not fit together very well, insert small strips of Evergreen strip stock into the voids and cover them with super glue. Then sand the surface of the plastic flat. Here again, replace the bolt head detail with Grandt Line detail bolts.

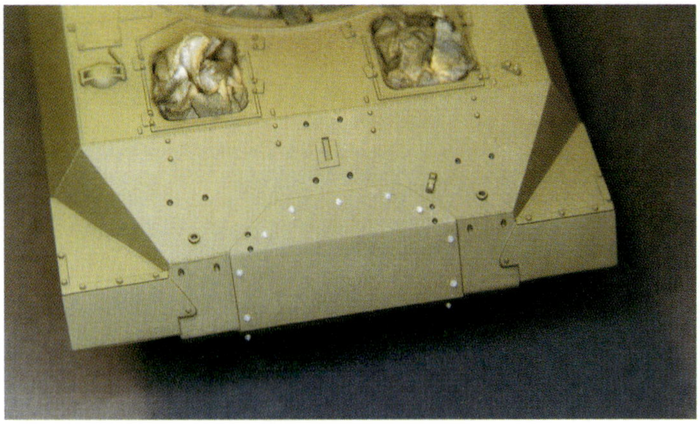

Install the bolt head detail on this M-18 Hellcat, and it is ready to be painted.

The back end of this tank did not fit together well, but liberal application of strips of masking tape helped position the parts and hold them in place while the super glue was applied along the seam lines.

It was not possible to fix the misalignment along the edge of this tank hull, but there is a simple solution.

Glue a piece of Evergreen strip stock into place as the first step in correcting the misalignment. When you glue the strip into place, be sure both contact surfaces are completely covered with glue so that there is a good seal.

Carefully shape the strip stock with sanding sticks in various grits.

Paint the repaired area, and the misalignment is not noticeable.

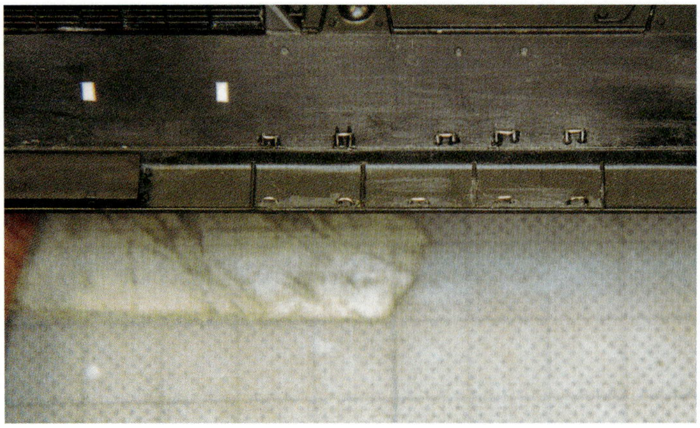

Sand off the molded details from this Russian T34/85 and replace them with wire. Note that they are not the same shape and are not equally spaced or positioned. Pictures of the real tank showed that these details were not uniform in either shape or location.

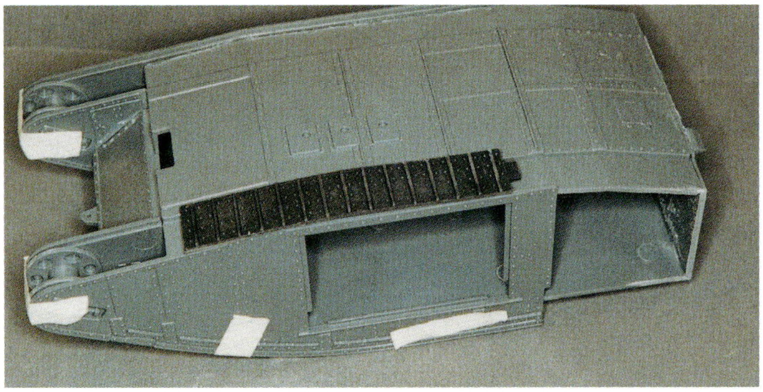

On this World War I female tank use the tracks to help position the tank's side panels correctly.

Install the exterior side panels along with the front inner panels. Note the seam between the panels. After painting the model, fill this seam with white glue.

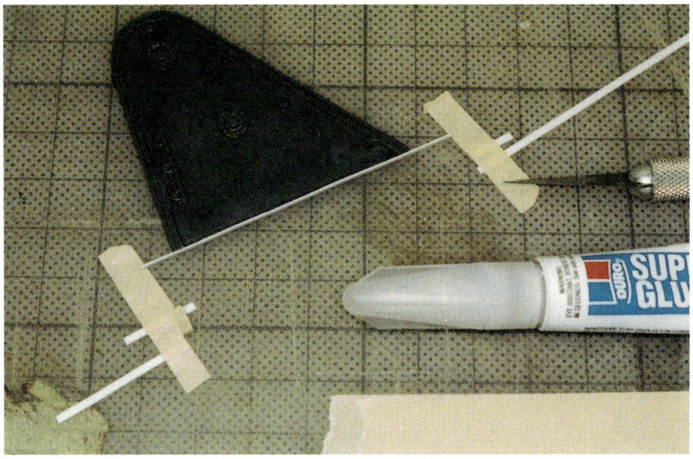

Since the rear inner panels do not line up evenly, add a piece of Evergreen strip to lengthen the part slightly. This problem became evident during test-fitting.

The two inner panels have been installed, but another problem became apparent that was unnoticed during the initial parts check.

The inner panels that sit across the bottom of the tank were too high. This presented a real challenge. In the end the solution was easy but messy. I carved and sanded down the high spots on the two inner panels. I checked and rechecked the progress of my work using a section of the track as a level check. Notice that that one side has already been cut down.

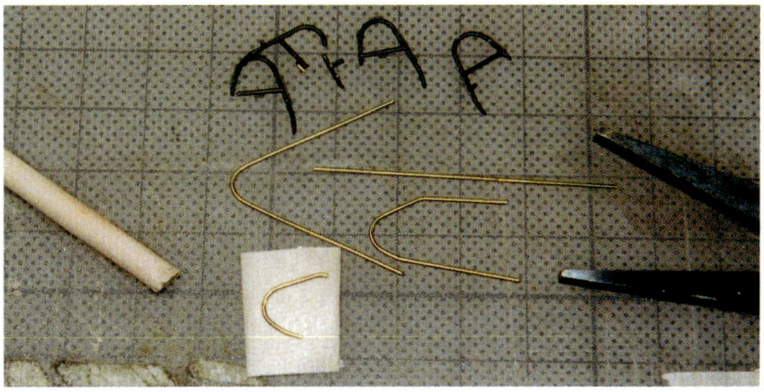

Even though I carefully removed the brush guards from the tree, they broke in several locations. Trying to glue them back together was just about impossible, so I made new guards from brass wire. To set the curves I used a wooden dowel, and to set the angles I used flat-faced pliers. Before bending the brass, sand the surface with 600-grit sandpaper.

Cut the cross bars from Evergreen rod stock and then glue them into place with super glue. Super glue bonds well to brass as long as the surface is clean and slightly roughed up. That's why you have to sand it with fine-grit sandpaper.

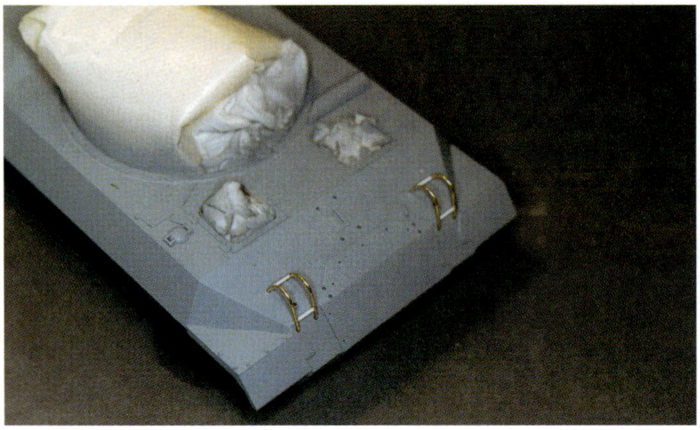

The brush guards are nearly complete, and they will look just as good as the kit-supplied ones. The entire scratchbuilding exercise took less than an hour. Learning basic scratchbuilding skills helps you fix these types of problems. It also encourages you to do more scratchbuilding.

After it has been painted, this Jagdpanzer IV hull is just about ready for its road wheels, return rollers, and individual track links. Note how clean this model looks. No seam or mold lines are apparent, and the paint job is superb. Model by Bill Teehan

This World War I female tank has been painted and the side decals added. Note that the seam line along the side panels has disappeared. Carefully apply white glue as a filler. Use several coats. Since you add the glue after painting the surface, you don't have to paint the glue. This is the great advantage of using white glue. Since it dries clear, it will transmit the color of the surface underneath it.

Install the hull and drive system on this M-18 Hellcat along with some of the hull fittings. Note how clean the running gear appears. Once the lower area on this model gets its dilute coats of mud and dirt, it will look more realistic.

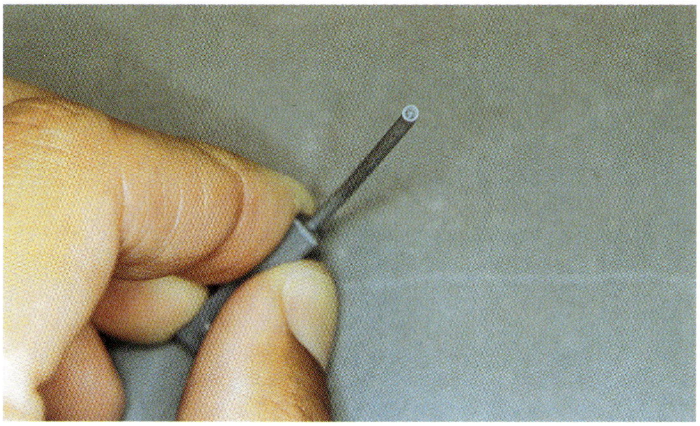

Small drill bits work well for drilling out an exhaust pipe. Paying attention to these little details can make a big difference in the overall appearance of your model.

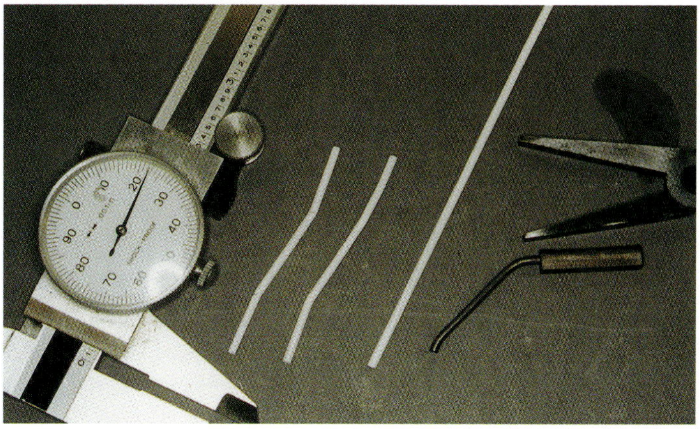

The exhaust piping between the exhaust collector and the muffler on this World War I female tank was not the correct length, so I made a new one using Evergreen rod stock. Here again, basic scratchbuilding skills and tools made the job easy.

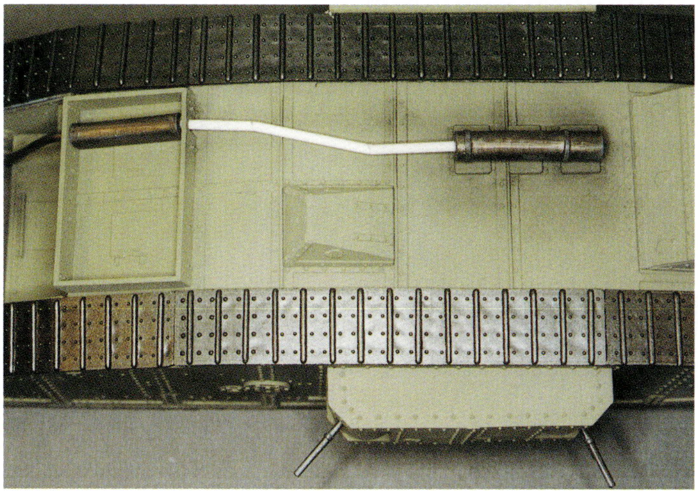

Here the tank's muffler pipe gets its final fit check.

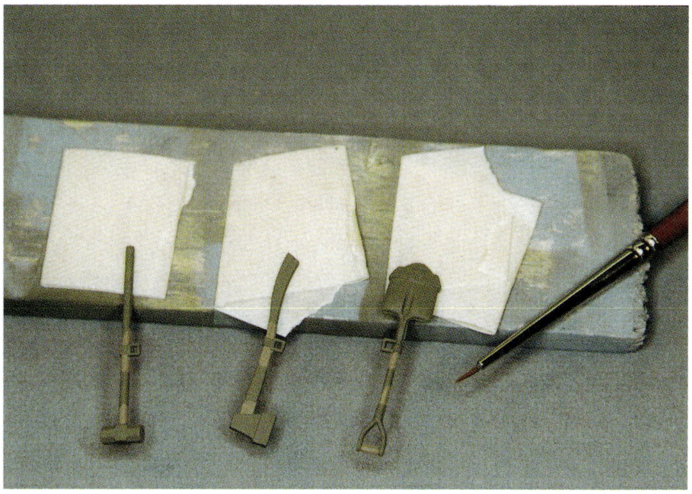

The best way to paint wood handles on tools is to outline the wood sections. Use a fine detail brush to set the demarcation lines between the wood and metal.

The next step is to paint the remaining sections of the wood handles with the same detail brush.

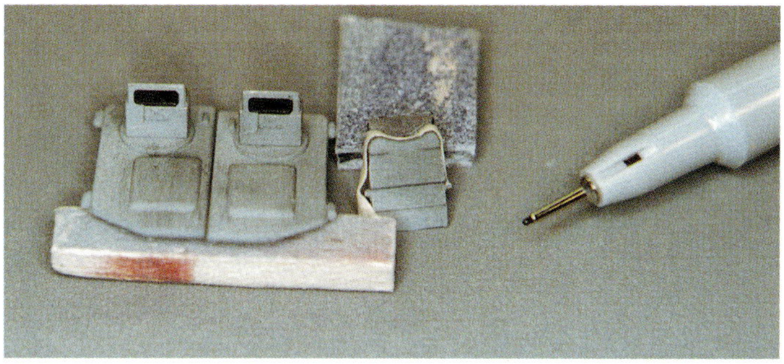

Fine-tip disposable drawing pens work well for highlighting the glass plating of these viewports.

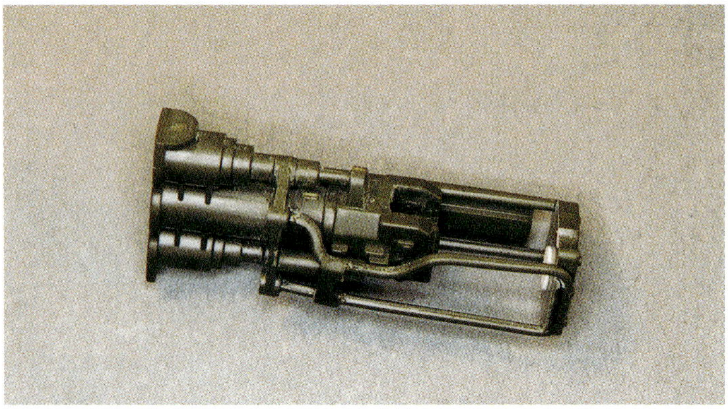

This breech assembly is now completely assembled and is ready to be painted.

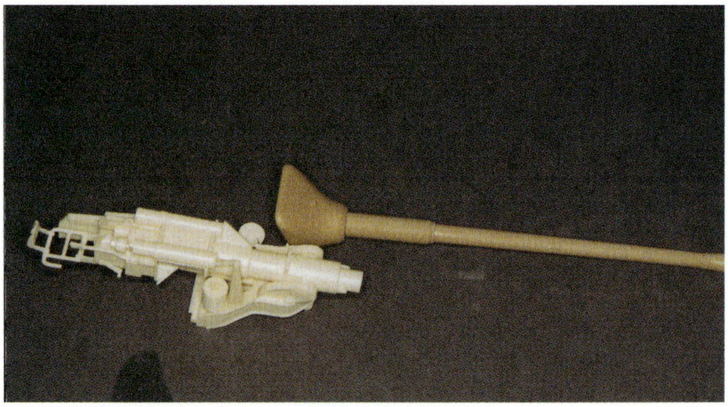

Here's another good example of a highly detailed resin kit breech assembly that is getting its final fit check on the kit-supplied mantlet and barrel. While these resin parts fit together well, they still needed some tweaking. Model by Scott Weller

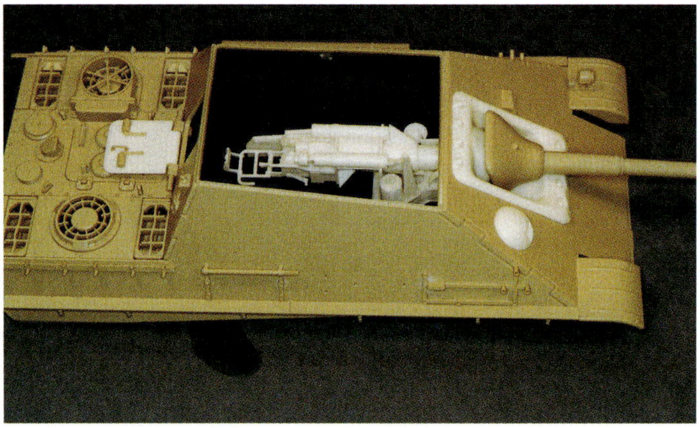

The resin breech assembly gets its final fit check to ensure that all the parts fit together correctly inside the front hull opening. Model by Scott Weller

Paint and then weather the breech assembly using pencil pastel dust to simulate grime, soot, smoke, and powder stains.

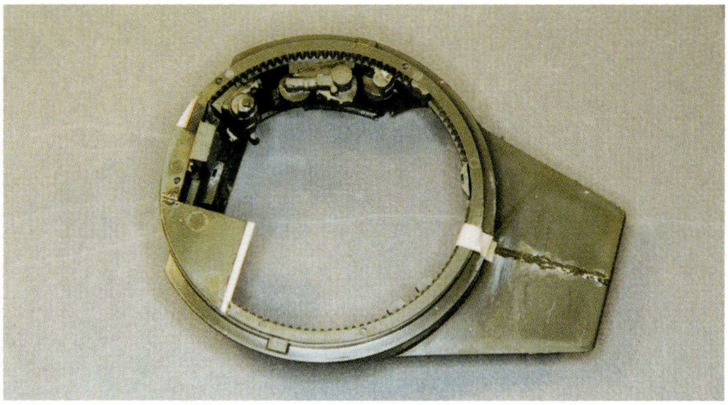

Assemble tank turrets that have inside detail parts slowly. As you glue the parts in place, check and recheck the fit to ensure that there are no clearance problems.

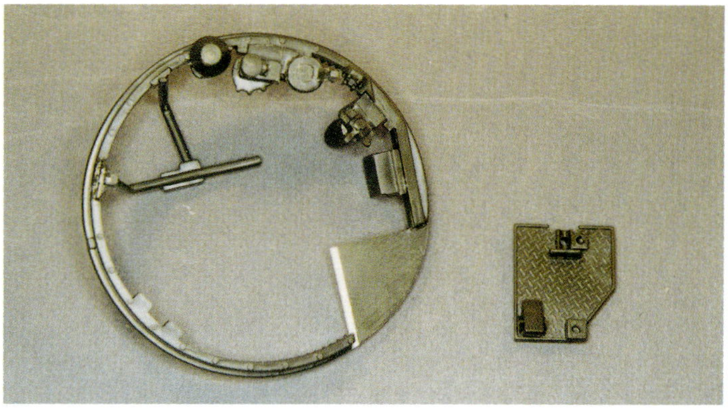

The inner turret ring and detail for this M-18 Hellcat is just about complete. Note the plastic strip, added to fix a fit problem.

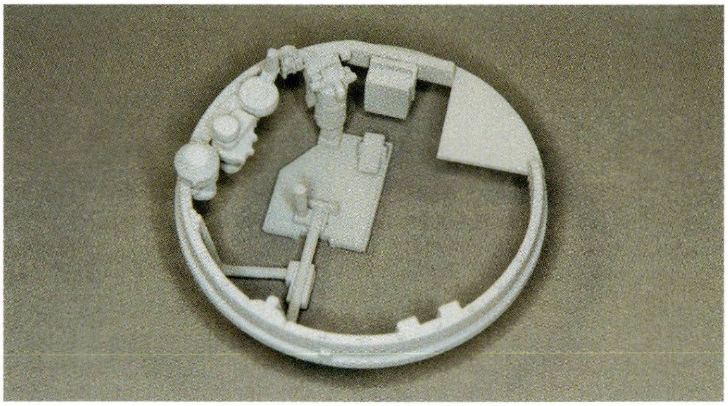

Here the inner ring has received its final coat of paint. The only thing left to do is weather it and add the seats.

Apply pastel pencil dust to the inner turret ring assembly. Note the streaky appearance of some of the dust. You can fix this by rubbing the streaky areas with a soft tissue. Also note that the decking where the crew members would stand received an extra-heavy coat of pastel dust.

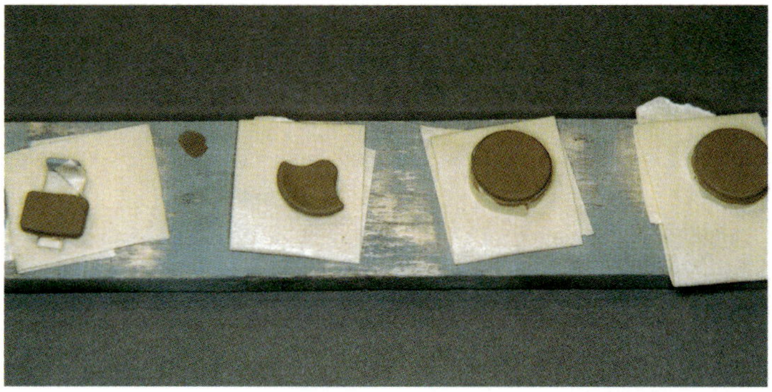

Note the masking around the seats to ensure a sharp demarcation line between the seat padding and the metal.

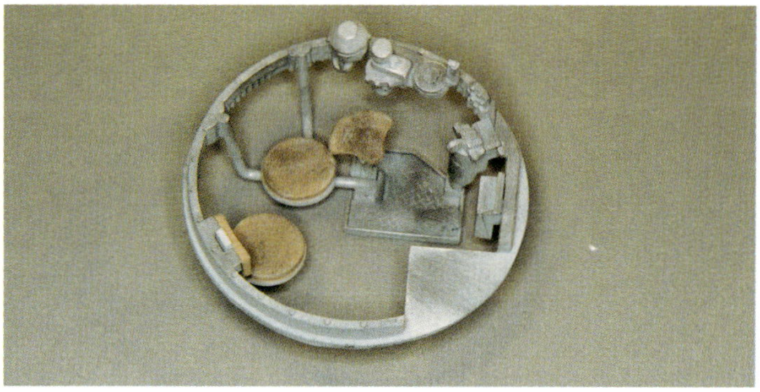

The complete inner turret ring assembly. Note that the color appears lighter and that the pastel pencil weathering does not appear as dark. Testors Dullcote lightens the appearance of paint and helps blend in the pastel dust. Drybrush the edges and corners and deck plating with Testors silver paint as well.

Tape together the left and right halves of the M-18 Hellcat turret and apply super glue along the seam lines.

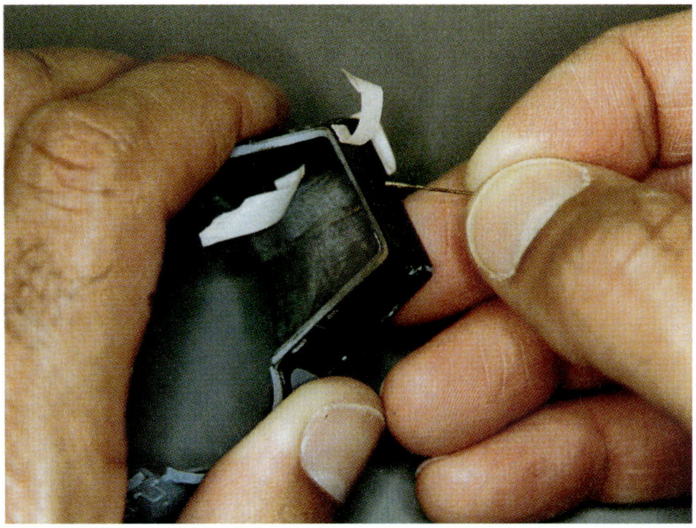

Once the glue is dry, remove the tape and apply glue to the remaining areas.

For smoothing out super glue on flat surfaces use a sanding stick. To smooth out the plastic and remove any remaining scratches, wet-sand the surface with 600-grit sandpaper.

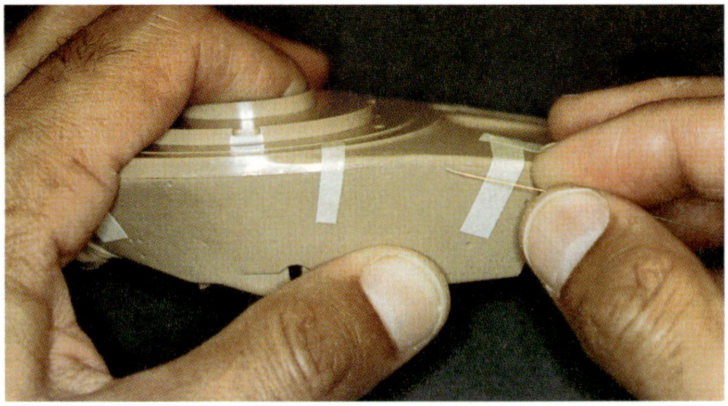

Glue the upper and lower halves of this M-60 turret the same way you glued the M-18 Hellcat turret. The difference here is that you must be careful when applying super glue along the seam lines of tanks that have casting detail on the turret surface.

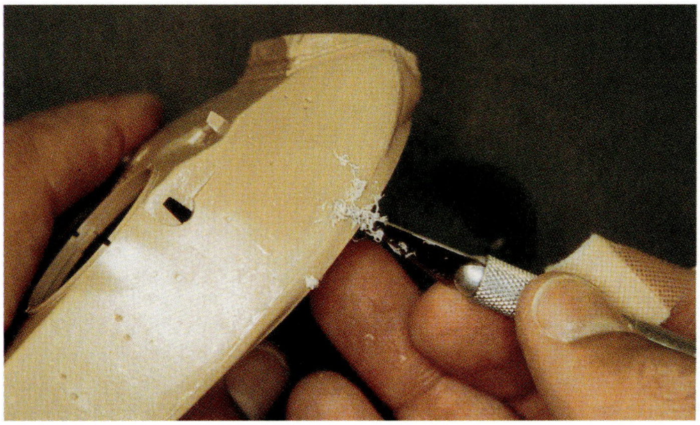

To avoid marring the cast surface detail, use the very tip of a number 11 X-acto blade to carefully scrape the super glue along the seam line.

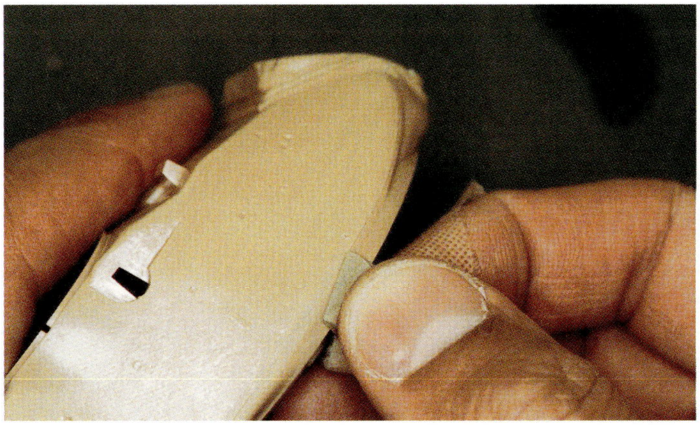

Use small strips of fine-grit sandpaper to smooth out the seam line and blend it into the surface of the plastic. Note that the sandpaper is touching the lower edge of the turret. Any damage to the surface casting detail will be concentrated on the lower area, where it will be less noticeable.

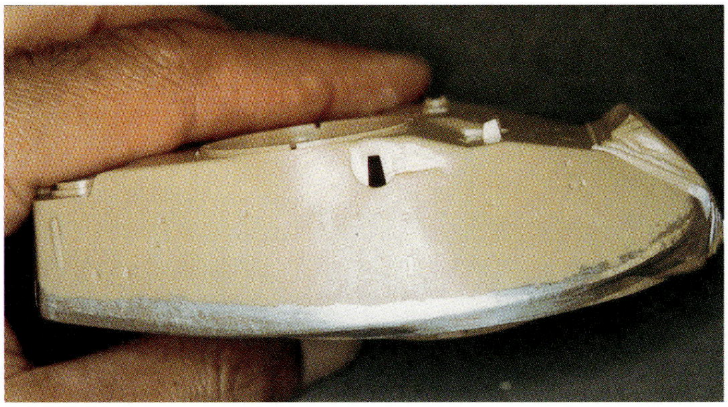

Apply Testors silver paint along the seam line. As you can see, very little of the casting surface detail has been damaged.

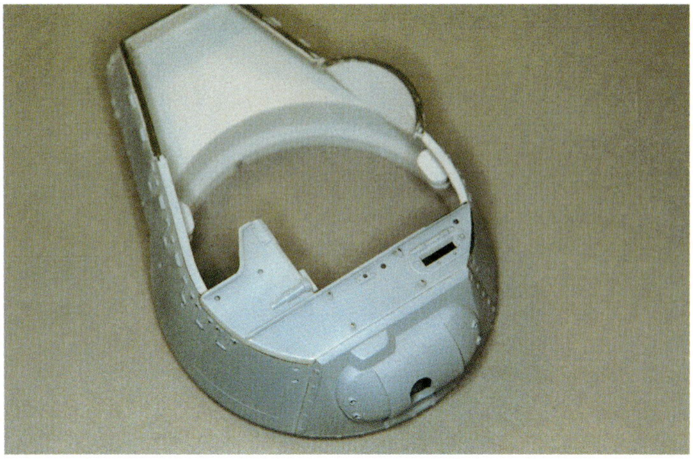

This M-18 Hellcat turret has been assembled and primed. The top of the forward plating has an uneven seam that should be fixed.

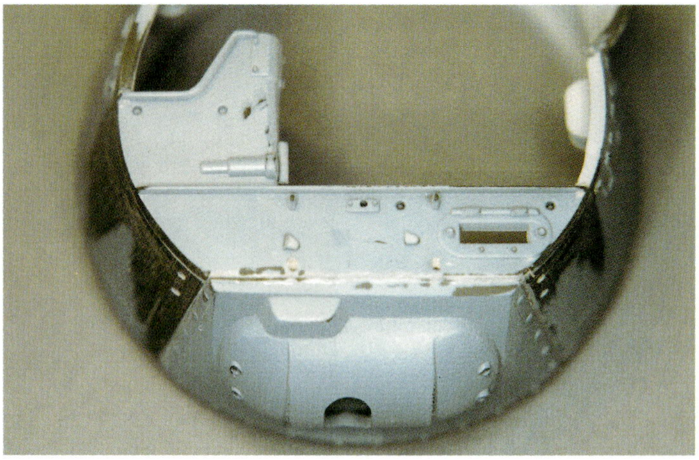

Carefully apply super glue along the seam line using a thin wire applicator. You'll also have to work on the seams along the left and right half of the top forward area of the turret.

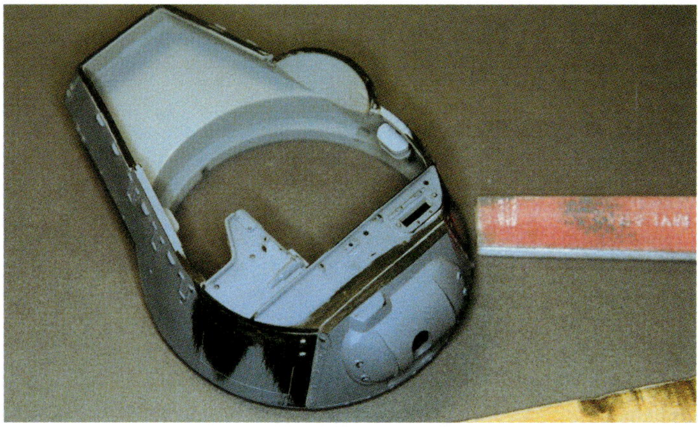

Sanding strips cut into thinner strips did the final sanding work to blend the super glue into the plastic. The thin strips made it easy to sand around the surface detail.

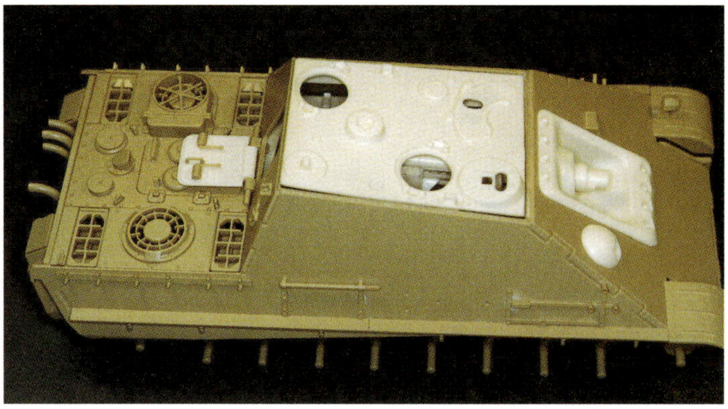

This Jagdpanzer is getting its final fit check on the new resin hull roof. The resin part is slightly warped. Fix this by gluing a thick length of brass wire to the underside of the part, where it will not be visible, to stiffen the resin and straighten it. Model by Scott Weller

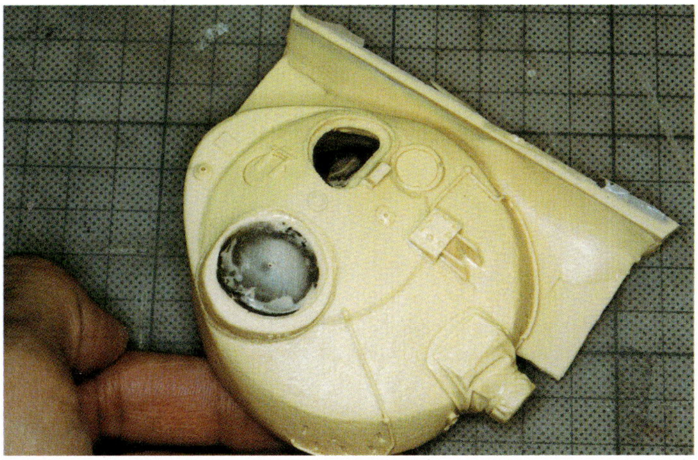

This Jaguar detail set turret has a rather large casting block. You'll have to remove it carefully.

Use a razor saw to cut away the casting block.

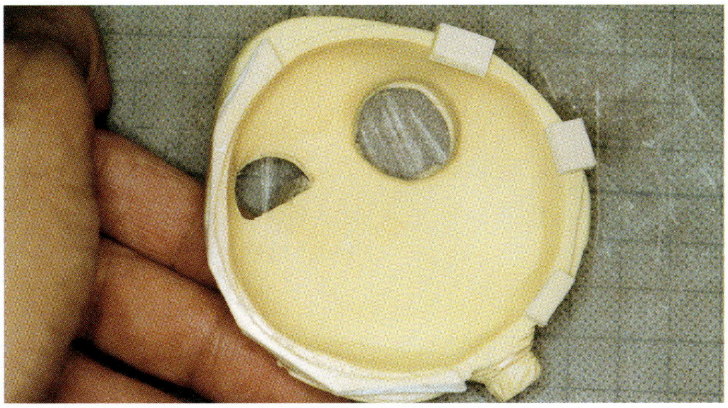

Since the casting block is on one half of the turret ring, it will be difficult to hold the turret level while you run it across a stationary piece of sandpaper to remove remaining resin. To keep the turret level, glue scrap resin strips on the other side of the turret ring. When you run the part across the sandpaper, the remaining casting block material will be removed evenly all around the turret.

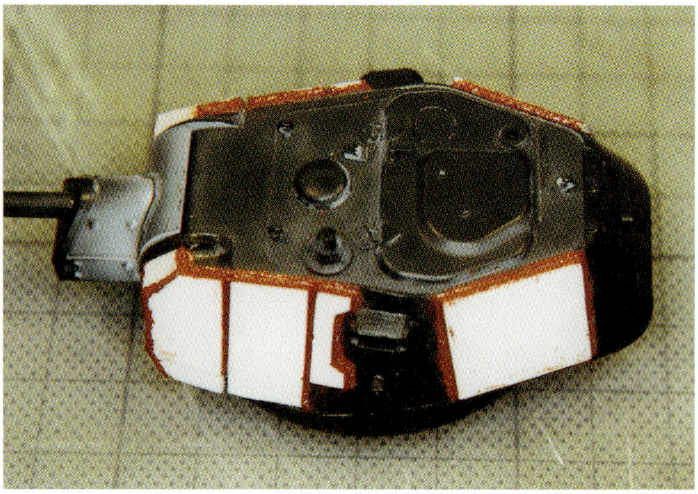

The part halves of these British female World War I tank turrets lined up correctly, but the gun port opening has a slight misalignment. To fix the problem, I carefully filed the curved parts of the opening with a round micro file until I had a smooth curve.

Add armor plating to this T34/85 tank turret using Evergreen sheet stock. Simulate the weld lines around the plating with automotive putty filler. Note that the plates are not all straight. Here again, reference pictures of the actual tank turrets showed that the plating was not always welded on straight. While this tank turret with its added plating may not be model-perfect, it does realistically portray what the actual tank turret looked like. Model by Jim Hudson

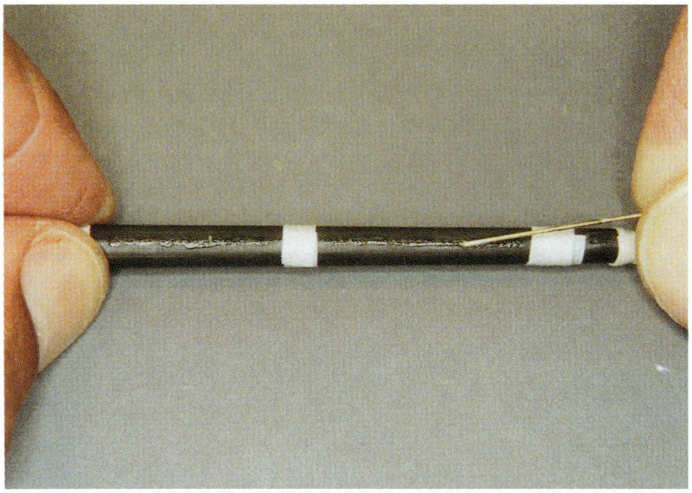

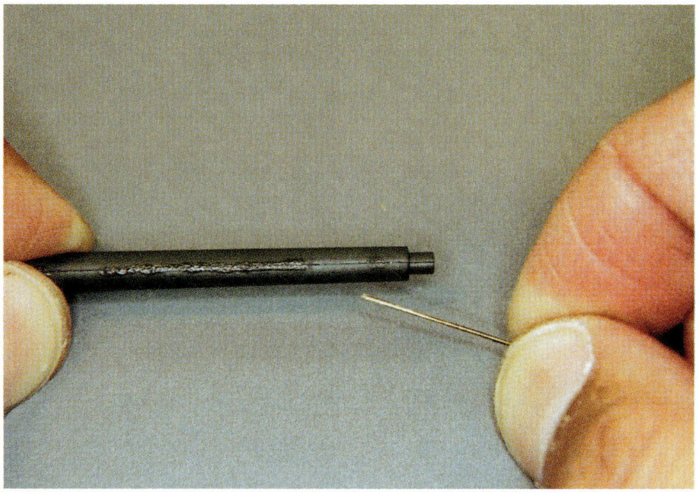

When gluing tank gun barrel halves together, use minimal amounts of glue. Remember to check the alignment of the barrel halves. If the alignment pins cause a problem, remove them.

Here again, use the same technique of positioning the parts, taping them together, applying super glue between the taped areas, and then removing the tape and finishing the job.

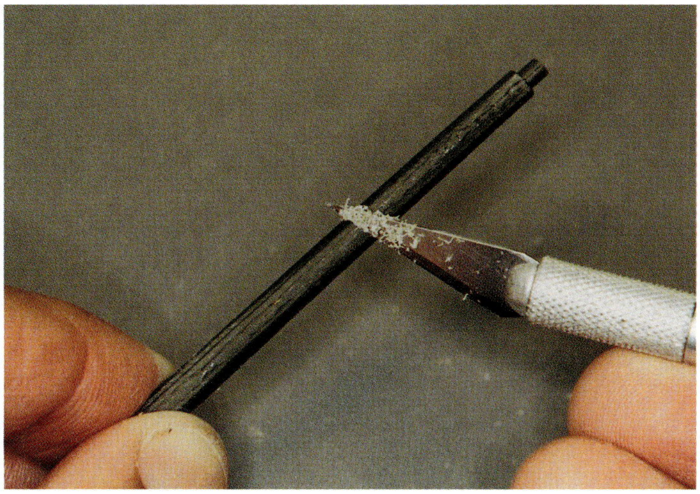

Tank barrels have small diameters, so be careful when you're scraping the super glue. You don't want to distort the shape of the barrel, which is easy to do.

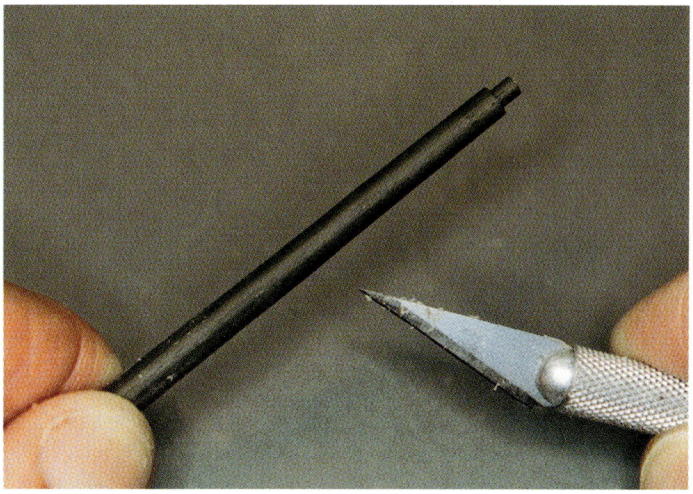

Gentle scraping with this sharp X-acto blade has carefully removed the excess glue without distorting the shape of the barrel.

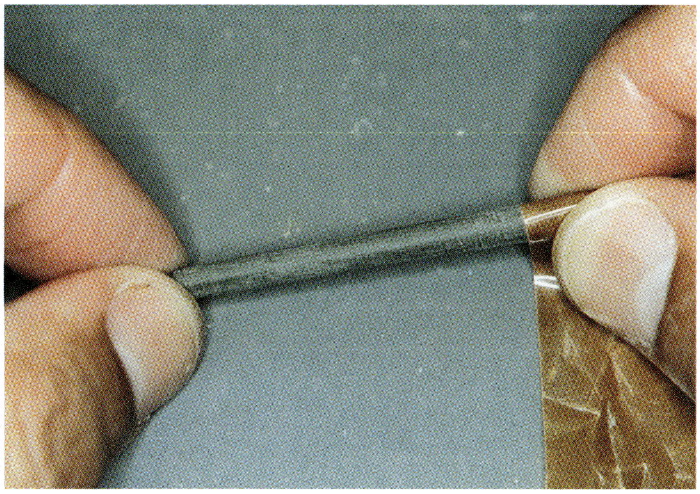

When sanding the seams on tank barrels don't sand along the seam line; you will create a flat spot along it. Instead, wrap the sandpaper around the barrel and push the barrel through the sandpaper with a twisting motion. This technique will ensure that the barrel maintains its round shape while blending the glue into the surface of the plastic. To polish the plastic use progressively finer grits of sandpaper and the same twisting motion.

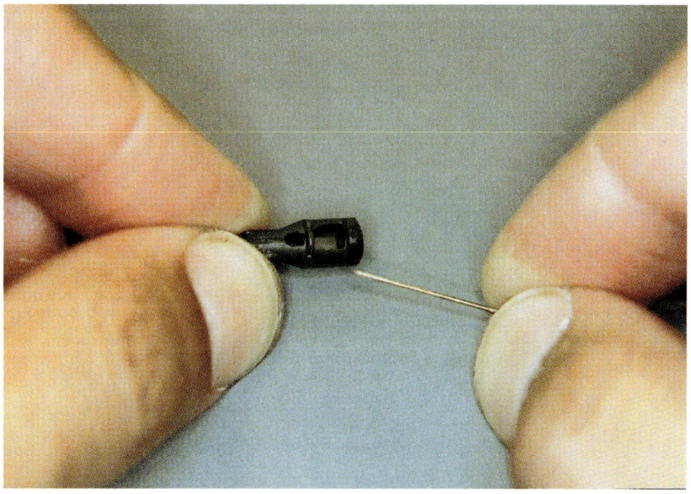

When gluing together muzzle brakes use very small amounts of super glue. It is easy to distort the shapes of the reinforcing ribs and blast deflectors on these parts with excess glue.

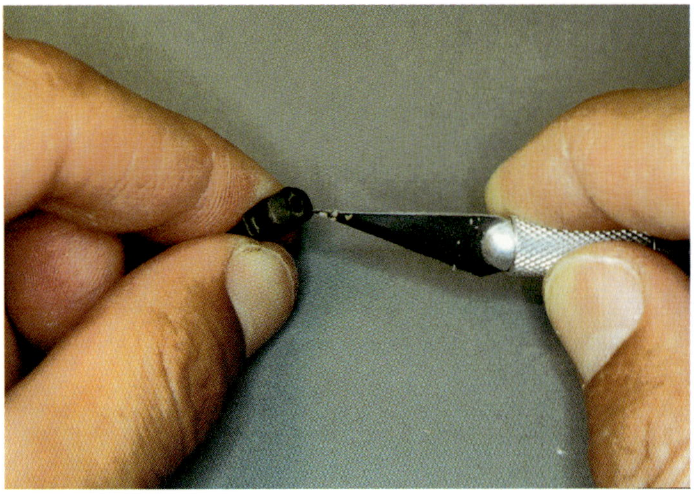

Here again, careful scraping to remove the surface glue ensures that you won't distort the shape of the plastic.

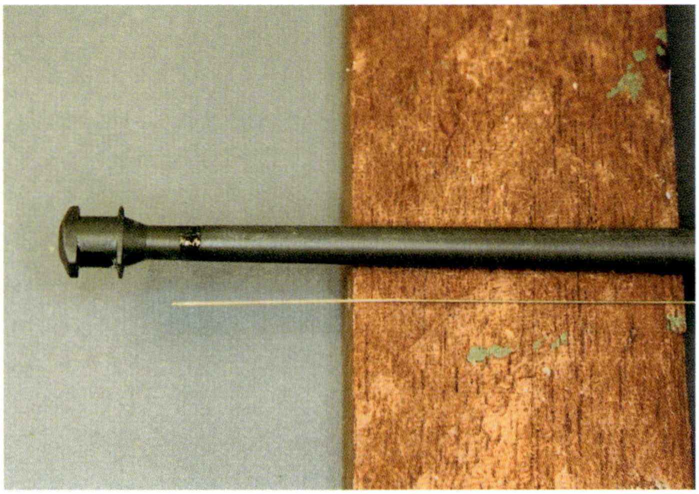

After gluing the muzzle brake in place, give the attachment point several coats of super glue. Dealing with this seam line in this location on the barrel will be a challenge.

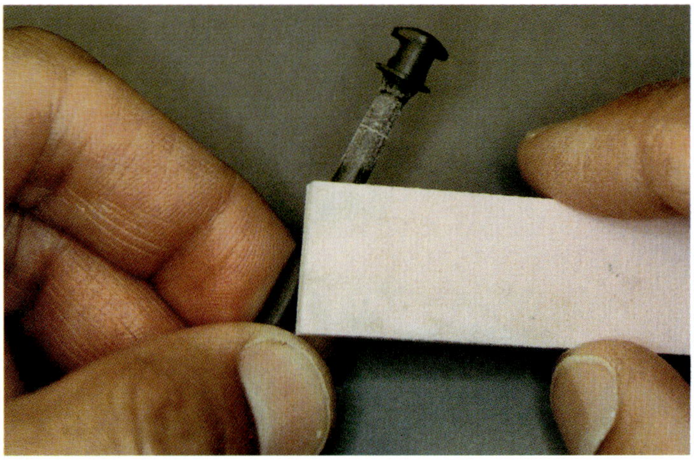

To maintain the roundness of the barrel, sand the super glue smooth and blend it into the plastic while rotating the barrel at the same time.

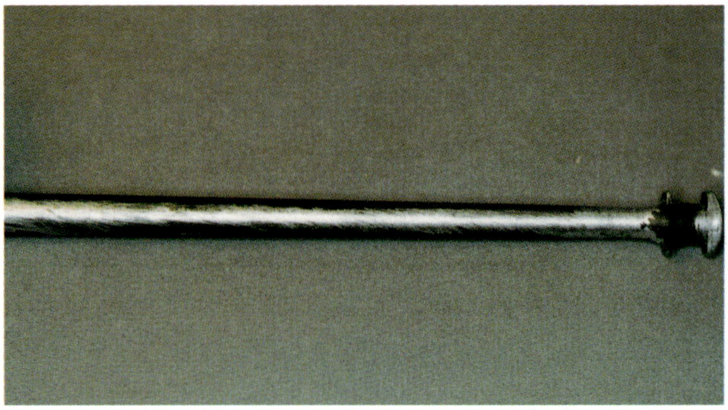

Use Testors silver paint to check the seam lines along the surface of this completed barrel, as well as to check for sanding scratches.

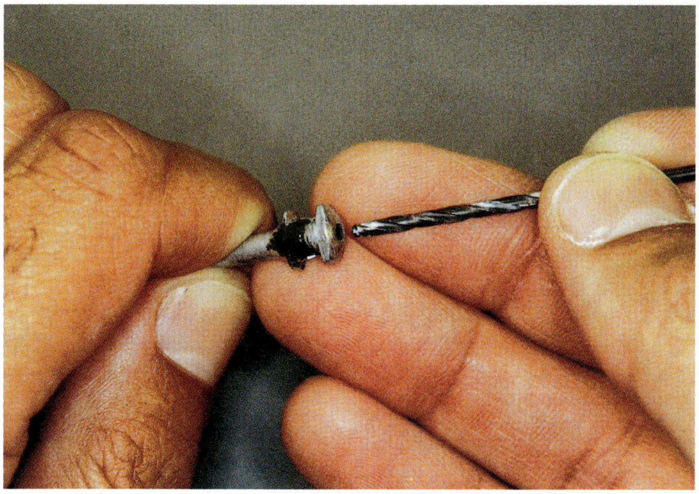

Use a drill bit slightly larger than the barrel opening on the muzzle brake to redrill the opening so that it has a nice round appearance.

As a final step apply Polly-S paint and decal remover to the barrel to remove the silver paint, wash the barrel to remove any residue, and polish the plastic with fine-grit sandpaper.

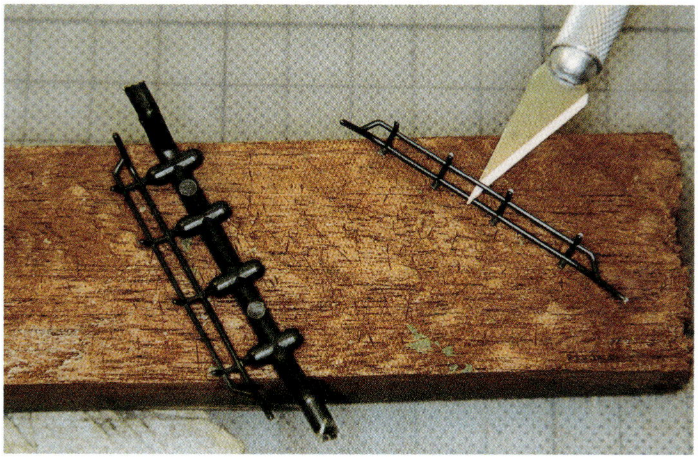

Here's another good example of how to remove delicate parts from their trees. Cut out the tree section holding the racks and then carefully cut off the parts.

The next step is to clean up the mold lines on these delicate parts. A number 11 X-acto blade in combination with fine-grit sanding sticks does the job.

White glue is great for attaching small parts.

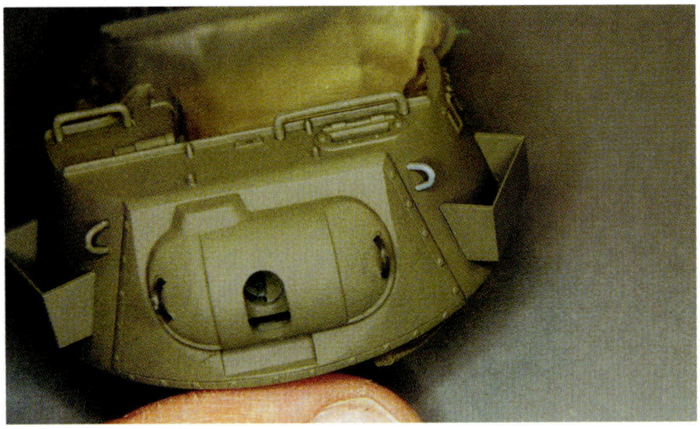

Use white glue to attach both lift rings and grab handles to this tank turret. White glue also fills the edges of the exterior storage boxes and seam lines.

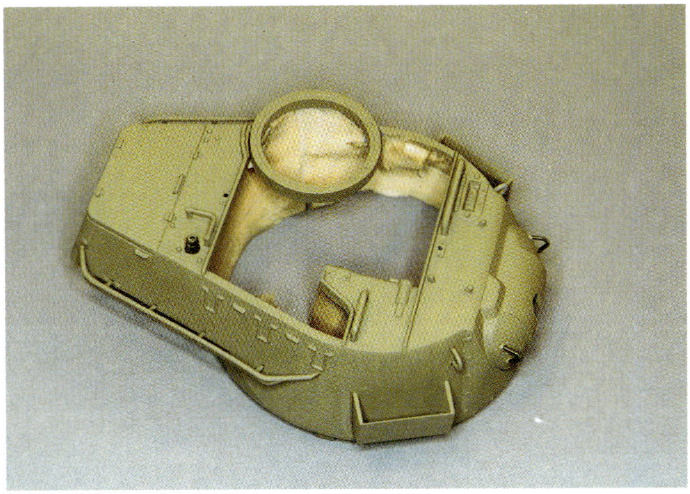

Mask the interior surface of this M-18 Hellcat turret so that there will be a fine demarcation between the interior gray and olive drab.

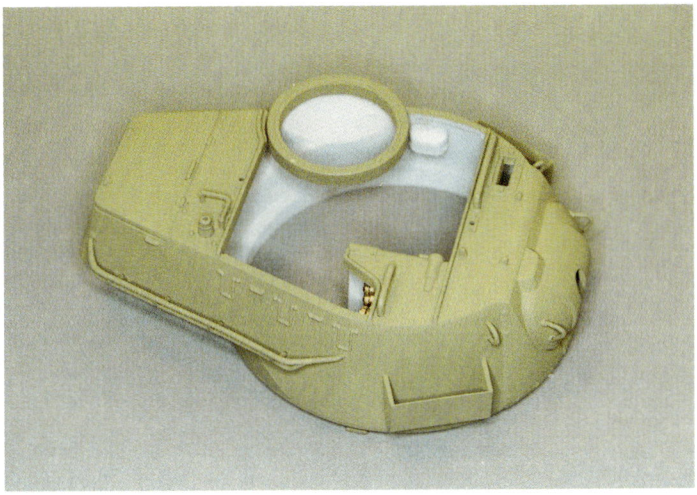

With the masking removed, you can readily see how a good masking job will result in a good paint job.

The interior walls of this tank also received a coat of pencil pastel dust, which was then sealed with Testors Dullcote.

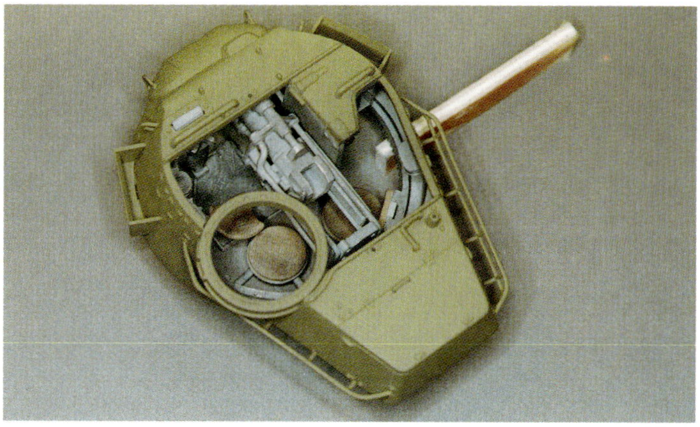

The breech for this M-18 Hellcat turret has now been installed. As you can see, it's getting crowded and busy inside this tank turret—just like the real thing!

With the gun barrel installed and the turret in place, this modeling project is almost done. Add the few remaining fittings and give the lower hull area its mud and dirt washes.

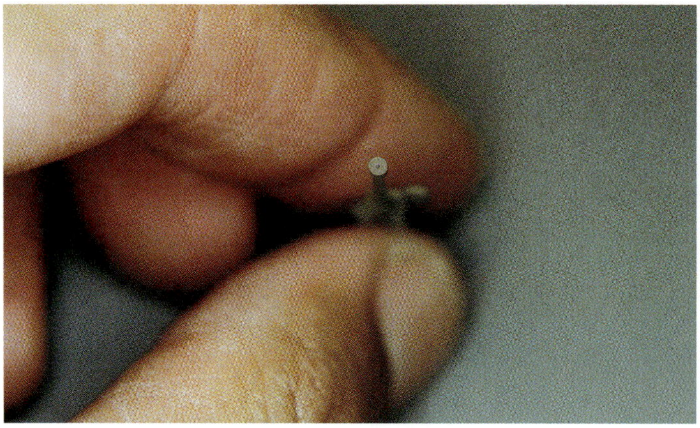

The first step in drilling out machine gun barrels is to indent the plastic surface so that the drill bit will not skip across the surface of the plastic. The indentation also allows you to check to see if it is off center. If so, just make a new indentation.

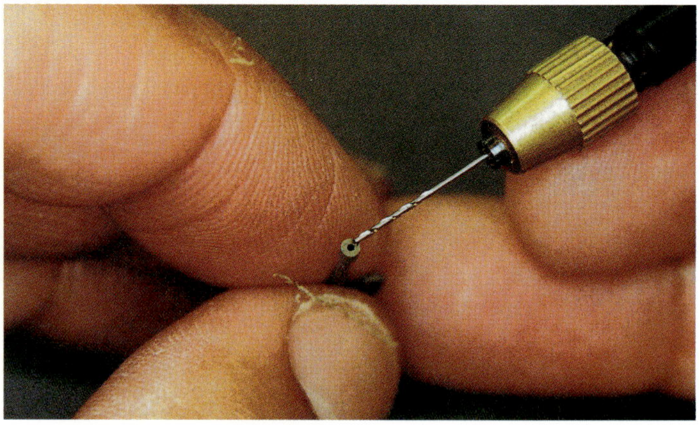

The next step is to use a small-diameter drill bit to drill a small hole into the plastic.

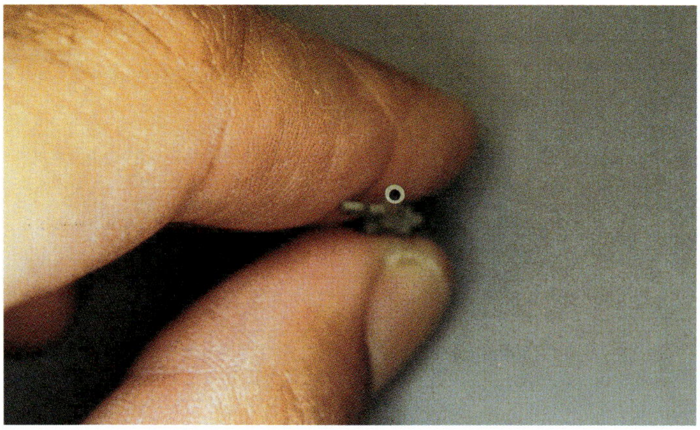

Next, use progressively larger bits to widen the hole.

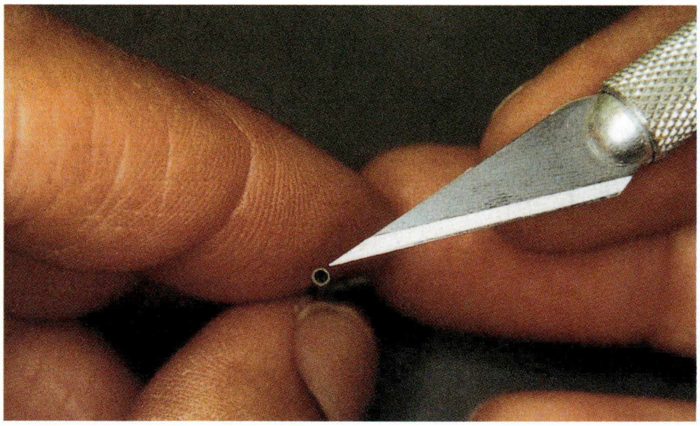

If you don't have a drill bit large enough to give you a thin wall around the opening, you can use the tip of a number 11 X-acto blade to carefully carve out the plastic. To do this, position the blade at an angle on the inside of the drilled hole and carefully rotate the part. The blade will peel away thin layers of plastic, widening the hole.

One last picture for this chapter. This British World War I female tank has received its mud and dirt weathering on both the tank and its tracks. Note that the front decal has also been weathered.

4 Building and Detailing Self-Propelled and Towed Artillery

Self-propelled artillery pieces are almost always built up on chassis designs similar to tanks, except that they do not have to carry heavy armor plating. The track systems are similar—there are sprockets, idlers, road wheels, return rollers, and the tracks. These are built up just like any other tank model, so refer to Chapter 3 for tips on how to assemble the chassis and track systems. The similarities between self-propelled artillery and towed artillery are the guns themselves, so let's start there and move onto the recoil, framing, and suspension systems.

Gun barrels. Building up gun barrels on artillery is similar to working on tanks, except that on large-caliber guns you do not have to be as careful about distorting the round shape of the barrel when you scrape and sand the seam area. Fit-check the part halves and remove any alignment pins that cause a problem. Sometimes the lower part of the barrel halves may be very slightly different in size. If you have this problem, try to work the misalignment into any area where you cannot see it. For example, if you remove all the alignment pins and the forward half of the barrel fits perfectly, there may still be a slight misalignment at the back end where the reinforcing rings are. This sometimes happens on long barrels. To fix this problem, align the rear top half—the bottom rear area will most likely be covered up by the slide. If I can cover up or hide a problem, I don't expend any time trying to correct it.

Tape the halves together with small strips of masking tape and then run a bead of super glue along the seam lines. Remove the tape and add glue to the remaining areas. Once the glue is dry, use a number 11 X-acto blade to scrape the glue flat carefully and gently. Here is where your fit checks and alignment efforts come in handy—you will not have to do much scraping, and the only sanding you'll have to do will be to smooth out the glue and blend it into the plastic. On large barrels I use sandpaper in combination with a Flex-I-File. The Flex-I-File is great for getting into the seam areas between the reinforcing rings of the lower barrel area. You can also cut the sanding strip of the Flex-I-File to a smaller width to get into those really tight areas. Like a Waldron punch tool, you will find a thousand and one uses for this simple but handy device.

Be careful about scraping and sanding around the muzzle brake area of the barrel. You don't want to distort the shape of the corners and edges. To check the seam areas, give the parts a coat of Testors silver paint, add more super glue if necessary, sand smooth, and remove the paint with Polly-S paint and decal remover. After using this paint remover, wash the parts in soap and water to remove the residue.

Once the barrel is complete, move onto the breech and slide assembly. Fix any imperfections on these parts, because they will be noticeable on the breech area. Be careful when cleaning up and gluing the interior ribbing of the breech. When assembling the slide, be sure to fill in the interior seam of the lower area, since this will stand out like a sore thumb. I use small amounts of filler applied with the tip of a stencil knife to fill in the seam. I then cut thin lengths of sanding sticks to get into these tight areas and sand the putty smooth by wet-sanding. Wet-sanding will prevent the sanding stick from getting clogged with putty residue. When cleaning up the teeth or gears on the bottom of the slide, use the tip of a number 11 X-acto blade. Sand the tooth surfaces with the edge of a sanding stick to remove scratches from your scraping and smooth out the surface.

Recoil systems. The recoil systems for large-caliber artillery pieces are nothing more than giant shock absorbers. Most are hydraulic, but some also have internal or external springs to absorb the recoil. For hydraulic systems, be sure that the small-diameter stem fits tightly into the face opening of the larger cylinder. If there is any space, you will need to glue a new surface onto the large cylinder face and drill out new openings so there is a tight fit. To simulate the polished metal surfaces of the small-diameter stem length that fits into the larger cylinder, paint the stem with Testors chrome color. Use a flat brush and rotate the stem as you paint it, so any noticeable brush strokes will simulate turned metal. Be sure the masking tape fits tightly around the diameter of the stem for a sharp demarcation line. To help ensure that the paint will not bleed under the masking tape, use minimal amounts of paint on the brush when you paint around the tape. Small-caliber towed artillery guns do not usually have recoil systems other than the gun slide, and some do not even have that.

Support framing. Support framing for large-caliber artillery depends on whether it is for a self-propelled gun or a towed gun. For self-propelled guns, the frame is nothing more than a carriage for the gun and slide assembly. The carriage contains the gears for elevation and rotation as well as the aiming mechanisms. Assembly is usually straightforward, but the manufacturers' instructions almost always depict inserting the slide assembly into the carriage as it is being assembled. I always assemble the carriage separately, fix any seam problems, and position the slide inside the carriage and let it snap into place. I have never had a problem pushing the upper wings of the carriage area slightly apart to slip the slide into place.

Support framing for large-caliber towed guns consists of the carriage as well as long rectangular shapes that are built up like a box. To help ensure that the framing will be straight and square, add small reinforcing strips of Evergreen strip stock to the interior of the framing as you build it up. Tape the four sides together with small strips of masking tape and continue to make adjustments to the parts until the framing fits together correctly. This process is repetitive, so be patient. Some manufacturers have lip guides along the interior edges of the framing to help position the parts against one another, but others do not. You can always glue small lengths of Evergreen strips to the inside of the part to create a positioning lip. When the parts are positioned, apply tiny amounts of super glue along one seam line between two parts. Once these two parts are dry, check the positioning of the remaining parts again and then apply super glue along the seam line on the other side of the part.

Now you have three sides glued into place. You can add the reinforcing plastic strips to the interior area and then finish the job. Small-caliber towed artillery framing is usually cast as either two halves with associated seam work or as one piece, so all you will have to do is remove the casting lines. When you scrape and sand, be careful not to distort the shape of the frame. Once the framing is completed, add details such as grab handles, racks, tool supports, and decals.

Suspension systems. Suspension systems for towed artillery range from complex axle, tie rod, leaf spring, and brake systems to just an axle and a set of wheels. When working with complex suspension systems, work from the inside out and be careful when removing the thin tie rods from their trees. It is very easy to break these parts. Cut around the trees that these parts are attached to so that you isolate the tie rod parts. Then gently cut the thin plastic connecting the part to the tree by placing the cut location directly over a hard surface such as a small block of wood. Be sure to use a sharp number 11 X-acto blade. Once you have completed the assembly of the suspension system, paint it, drybrush the corners and edges, and then add the wheels.

Weathering. In all the photos I have seen of self-propelled and towed artillery, the small-caliber artillery pieces always seem to look more weathered and used. Maybe it's because they are more easily moved by smaller vehicles, so they get a lot more abuse. This may also be because smaller-caliber artillery has a much shorter range, and consequently they are on the front lines while their larger brothers can lob projectiles from a considerable distance away from the front. Whatever the reason, your weathering should reflect this. Smaller-caliber artillery should get heavier coats of mud and dirt on the frames, suspension systems, and tires.

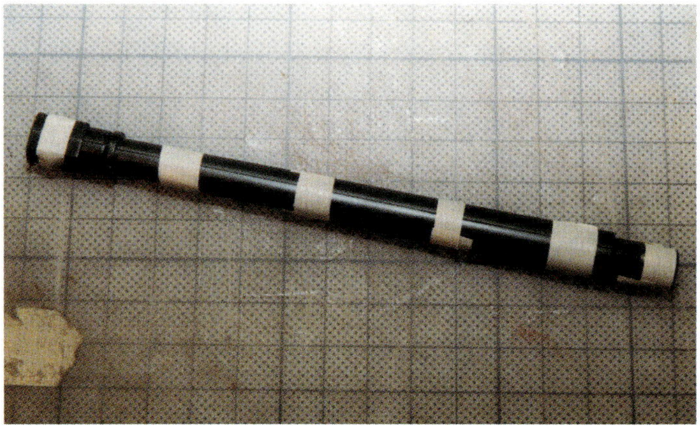

Assemble and glue big-diameter gun barrels the same way as little ones. Always check the fit first, especially around the reinforcing rings.

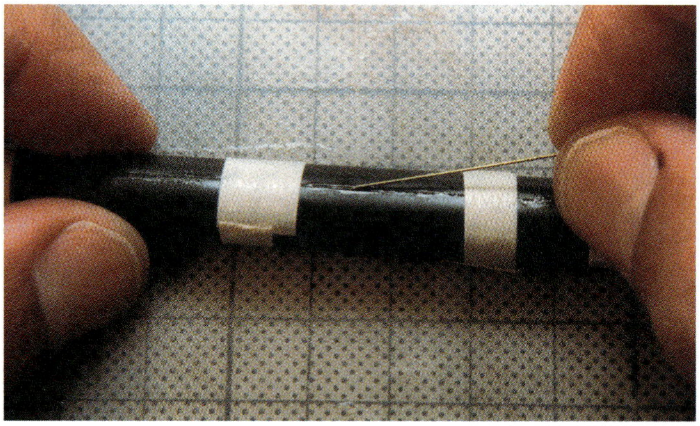

Here again, use small amounts of super glue to seal the seam line.

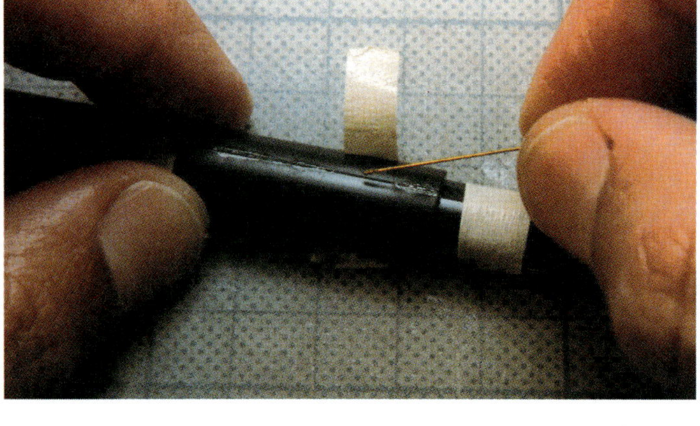

The opening at this seam location will need several application of super glue to fill it in.

While you still need to be careful when scraping and sanding large-diameter barrels, distorting their round shape is less of a problem.

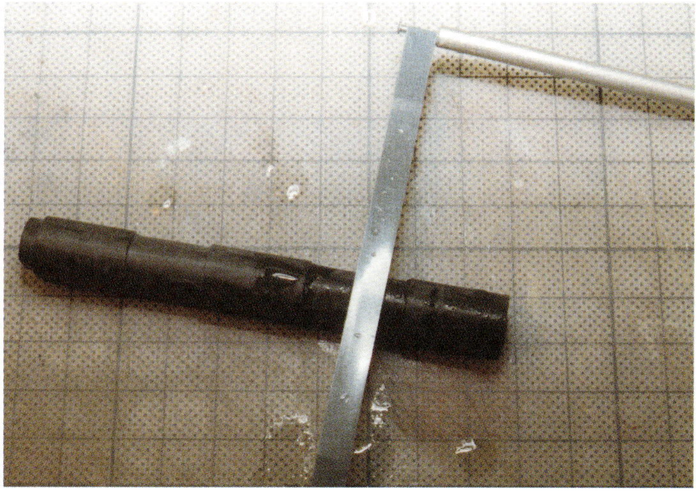

A Flex-I-File comes in handy when you need to sand around gun barrel reinforcing rings.

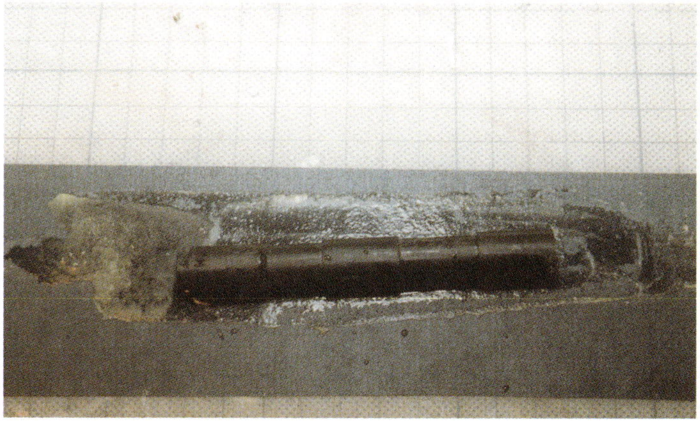

To sand smooth the flat bottom of the barrel, run it across a stationary piece of sandpaper. Sometimes I wet-sand to polish, and sometimes I wet-sand to prevent the sandpaper from getting clogged with sanding dust.

The end of this barrel was not flat because of a slight misalignment, but here again the solution is simple. It takes only a few passes across the sandpaper to fix these types of problems. Be sure to rotate the part so that you get an even surface.

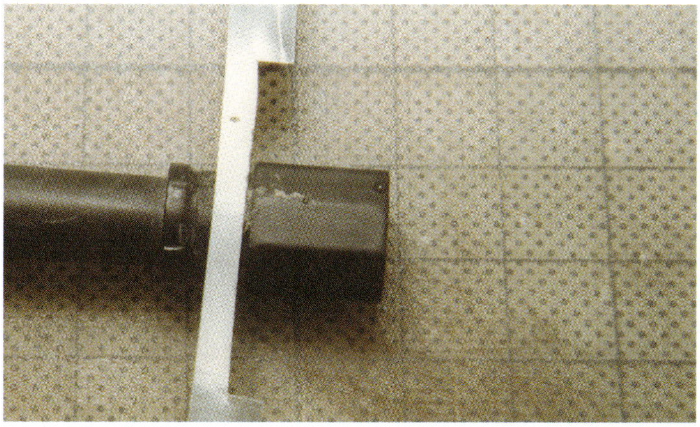

To get into tight locations, simply cut the Flex-I-File sandpaper strip.

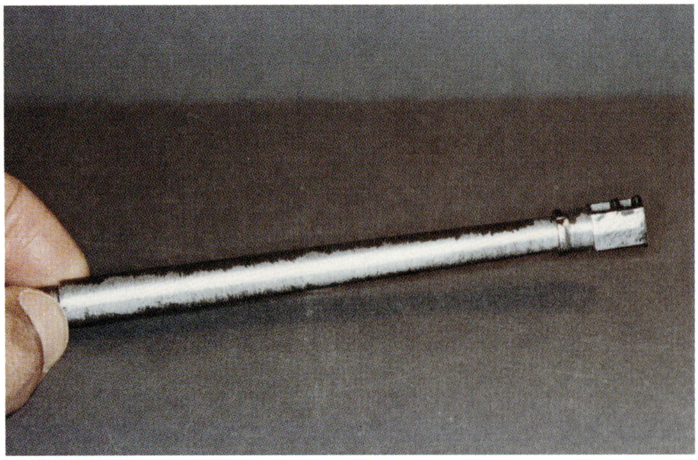

Once again, use Testors silver paint to check for any remaining flaws.

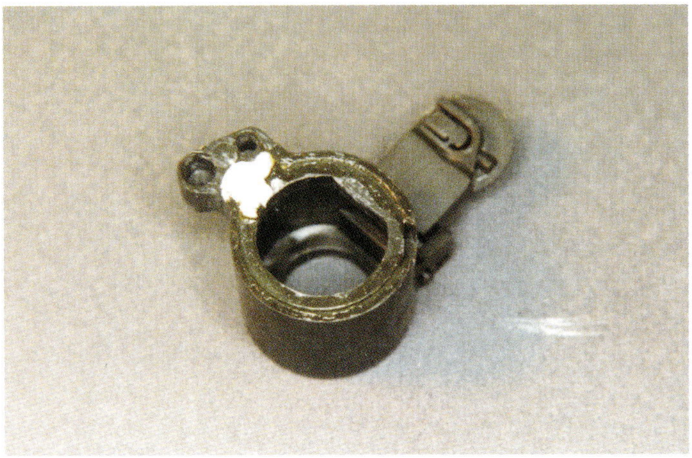

To fix the seamline and sink mark on the back side of this breech assembly, use a combination of super glue and Testors putty.

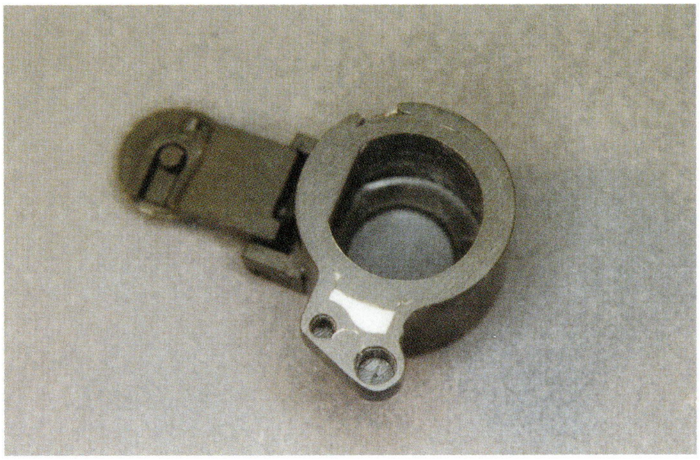

Sand smooth the face of the breech by running it across a stationary piece of sandpaper.

This breech assembly is ready to be attached to the barrel. The ribbing adds an extra level of detail to this model.

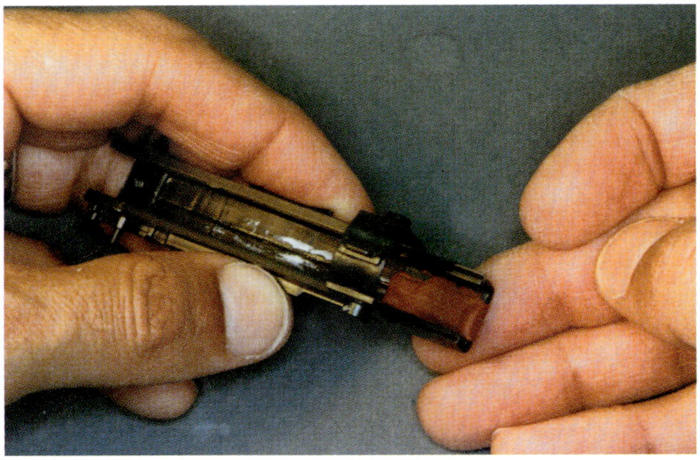

The interior seam area of slide assemblies can be hard to get to. The only area you need to be concerned with is what will be seen. Apply automotive filler with a flat-faced X-acto blade.

Sand the putty smooth with sanding sticks cut into thin lengths. To cut the sanding sticks, use a single-edged razor blade.

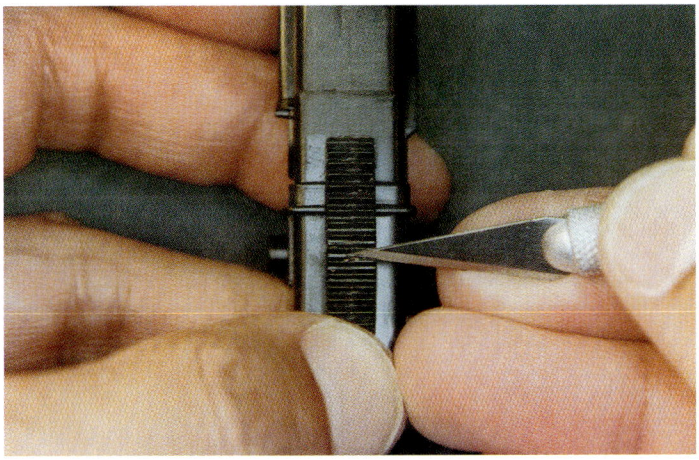

Be careful when scraping seam or mold lines on gears.

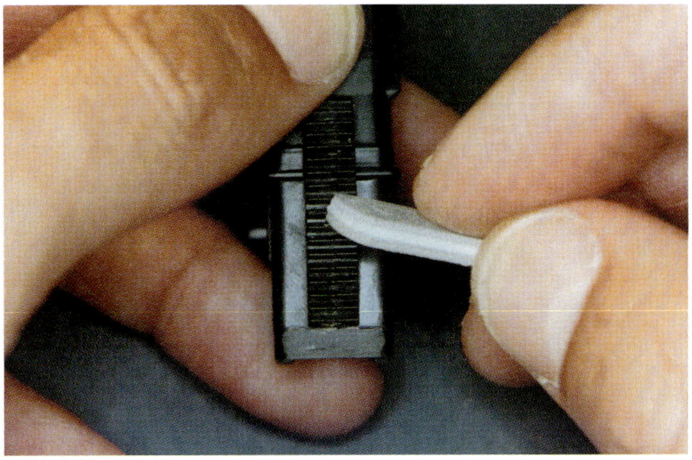

Use the edge of a sanding stick to smooth the plastic after scraping the gears.

tips

- Use Testors putty or Evergreen plastic strips to hide sink marks in locations that are hard to get to.
- To get into tight places, cut your sanding sticks into thinner lengths using a single-edged razor blade.
- Use a Flex-I-File to help shape gun barrel reinforcing rings.

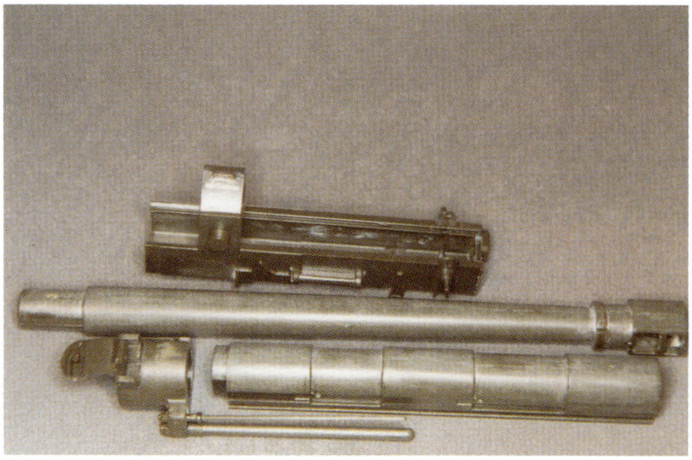

The completed subassemblies on this Italeri M-110 are now ready for priming and painting.

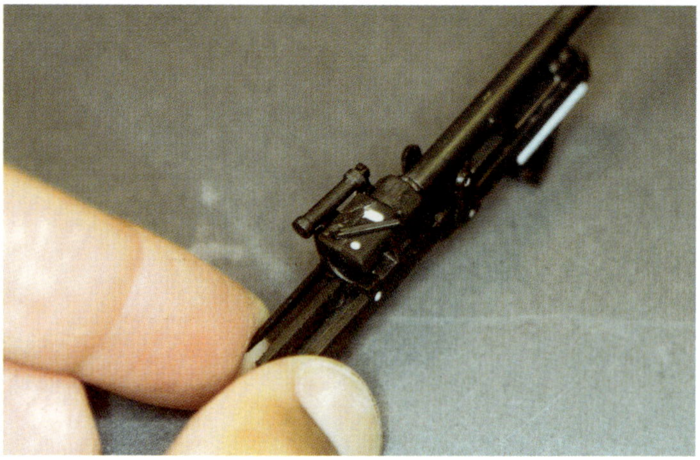

The breech assembly on this German 75mm antitank gun had injection marks and dimples that were in very tight locations. The solution was to cover them with small bits of Evergreen stock and disks punched out with a Waldron punch tool.

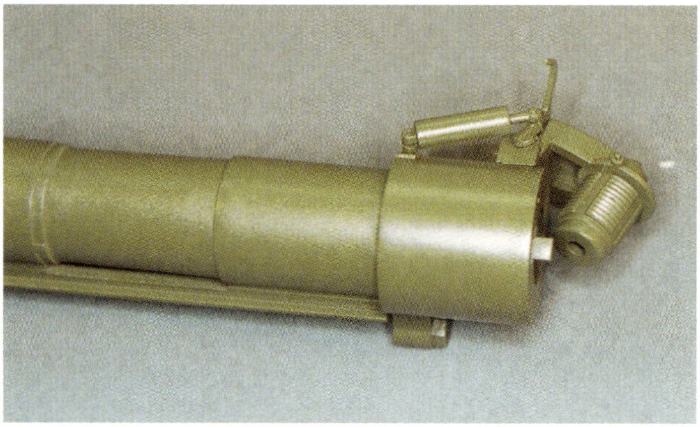

The painted barrel and breech assembly on this ARV Club 155mm Long Tom have been painted, and it is ready to be slid into the carriage assembly. Note how clean the breech assembly looks. Model by Richard Boutin, Sr.

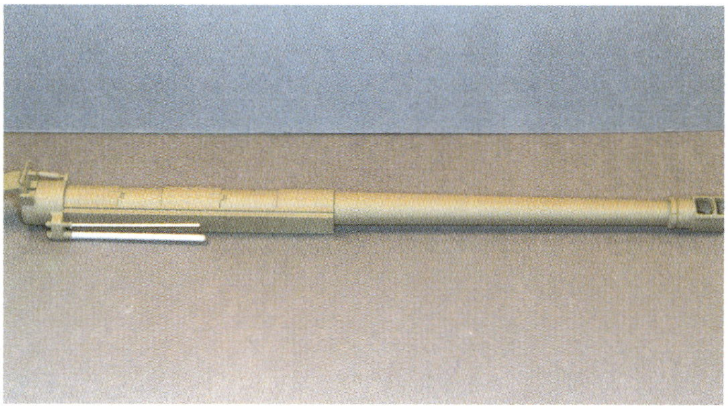

Don't forget to paint the hydraulic cylinder stems on the underside of the slide assembly.

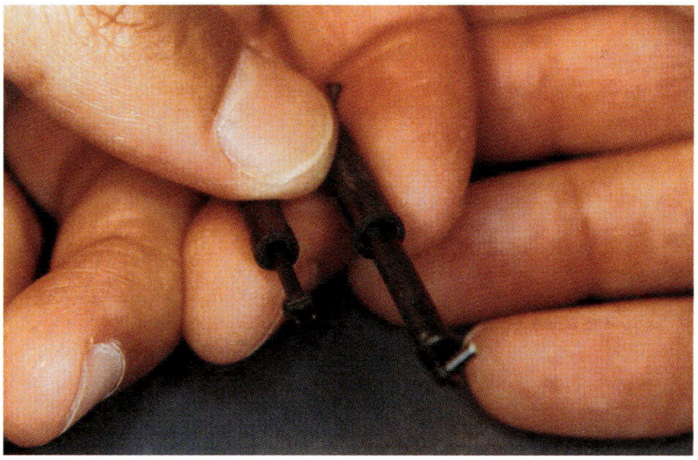

The openings on the faces of these hydraulic cylinders are too big for the stems. Once they are painted, the finished assembly would not be accurate or realistic—the fluid inside the cylinders would pour out. To fix this problem, you'll have to attach new faces to the cylinders.

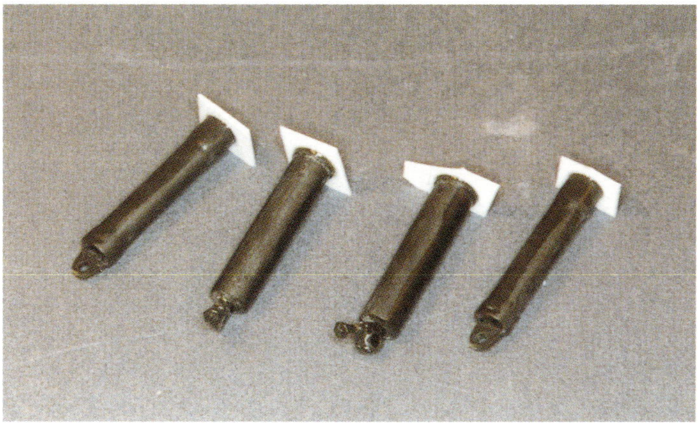

Glue scraps of Evergreen sheet stock to the faces of the cylinders. Be sure that the gluing surfaces are completely coated with super glue. Since you will be drilling through the sheet stock, use .020"-thick stock.

Use sharp blades to make tangential cuts on the sheet stock around the rim of the cylinder face.

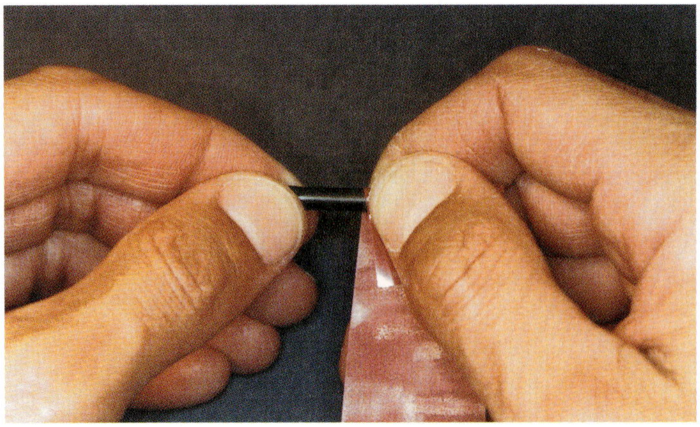

To round out the new face and blend it into the edge of the cylinder, wrap sandpaper around the end of the cylinder and then twist the part.

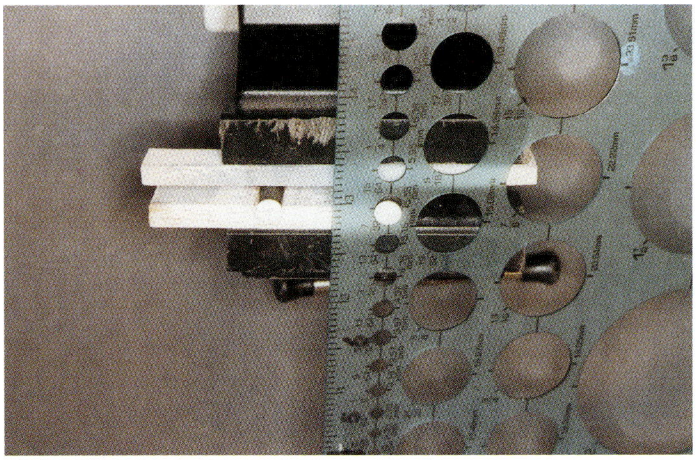

Next, you'll have to locate the center of the new face for drilling. I used a drafting circle template to locate the center.

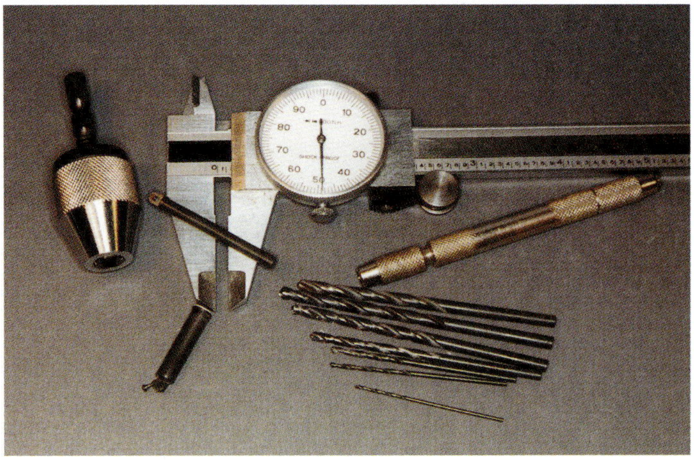

I then selected the drill bits I planned to use. I measured the diameter of the stem using a caliper and selected several drill bits so that the hole would get progressively bigger.

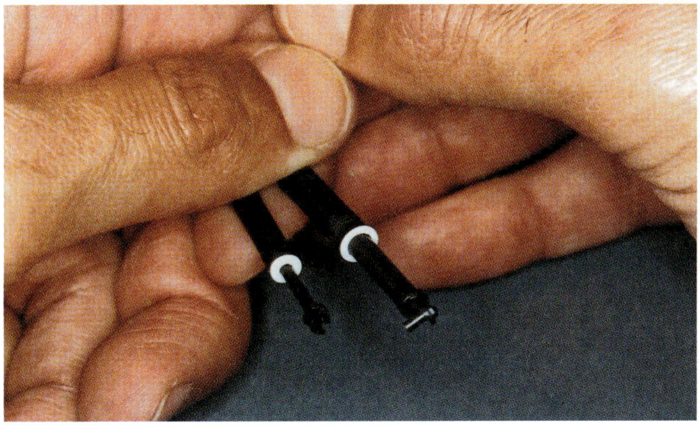

The stems now have a nice tight fit inside the opening of the cylinder faces.

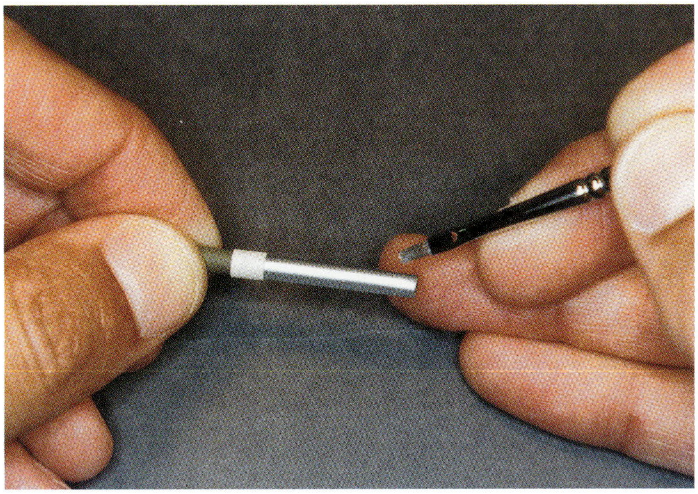

To set the demarcation line on the stems, wrap masking tape tightly around the stem. I use Testors chrome color applied with a small flat brush to simulate polished metal. To simulate the turned appearance of the metal, rotate the stem and apply the paint at a right angle to the stem. The surface of the chrome paint will have a subtle streaky appearance that simulates turned metal.

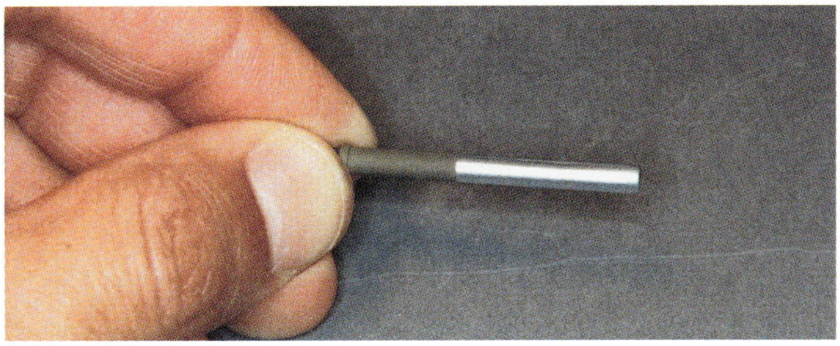

The finished stem has a sharp demarcation line.

Careful positioning and taping of the cradle halves will ensure a minimal amount of seam work along the perimeter of these subassemblies.

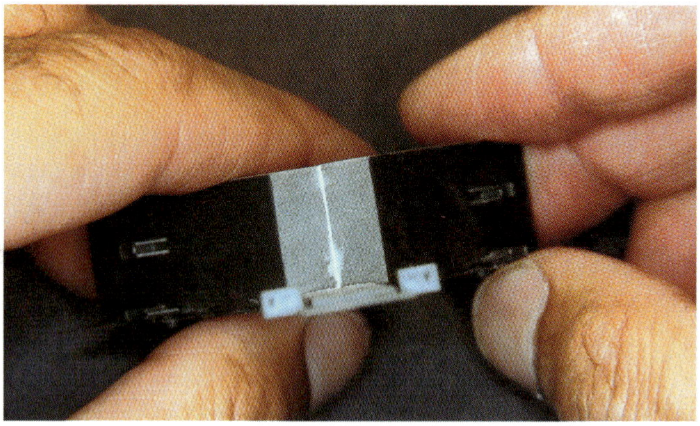

This cradle base is an excellent example of good seam work. Model by Richard Boutin, Sr.

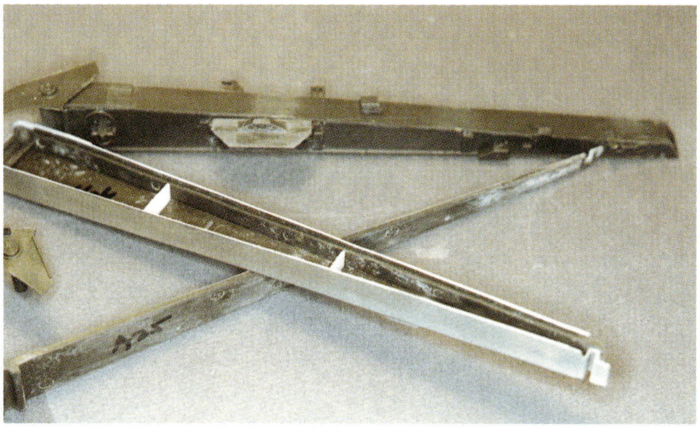

Evergreen strips help to build up and position the long rectangular panels of framing on this ARV Club Long Tom. Model by Richard Boutin, Sr.

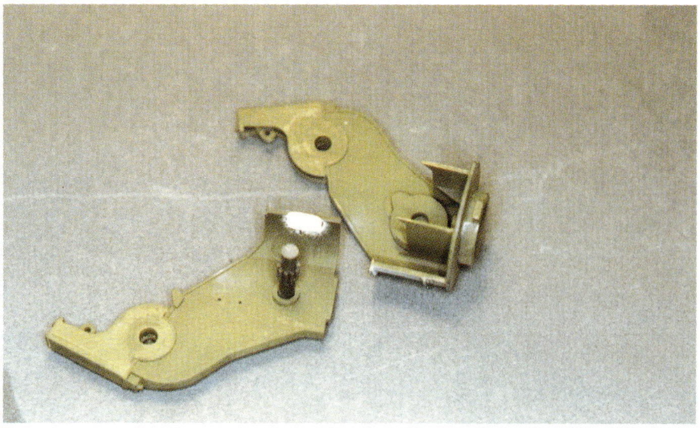

A small piece of sheet stock added to the inside lip area of this Long Tom cradle helps increase the strength of the gluing surface. Model by Richard Boutin, Sr.

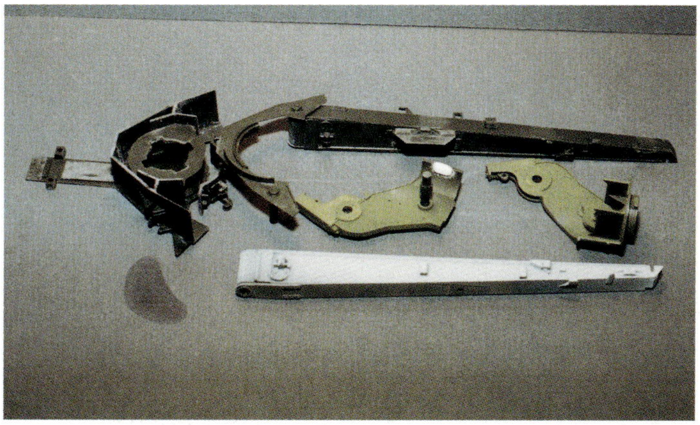

The cradle base and framing on this ARV Long Tom is starting to take shape. Model by Richard Boutin, Sr.

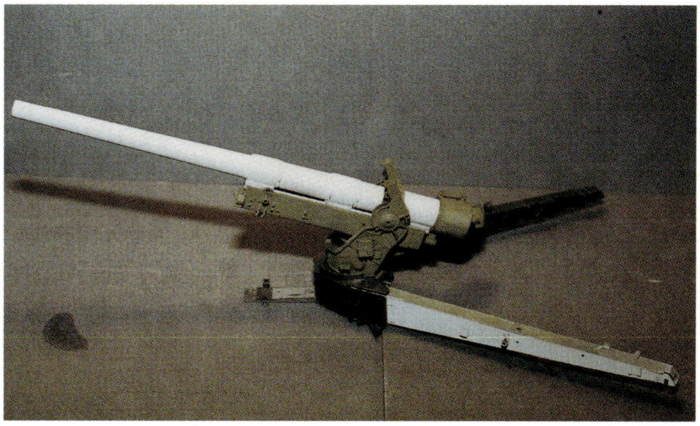

This is one of those exciting moments in scale modeling—a test-fitting of all the major subassemblies. Model by Richard Boutin, Sr.

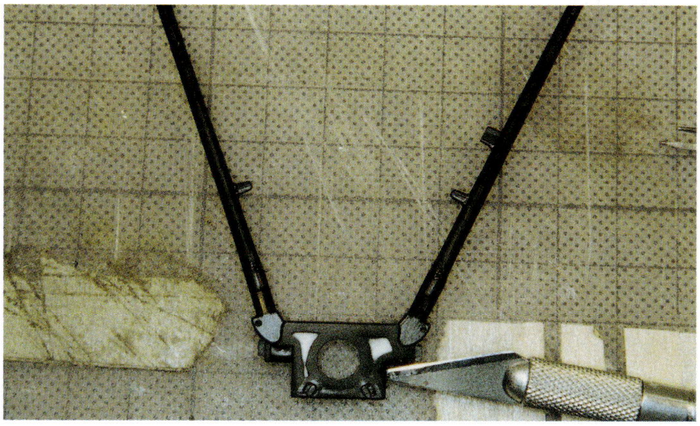

Test-fitting and assembly on small-caliber towed guns is pretty straightforward. Typically, the base and framing are single pieces that do not require a lot of assembly. The real challenge of this kit was dealing with the sink holes and injection marks.

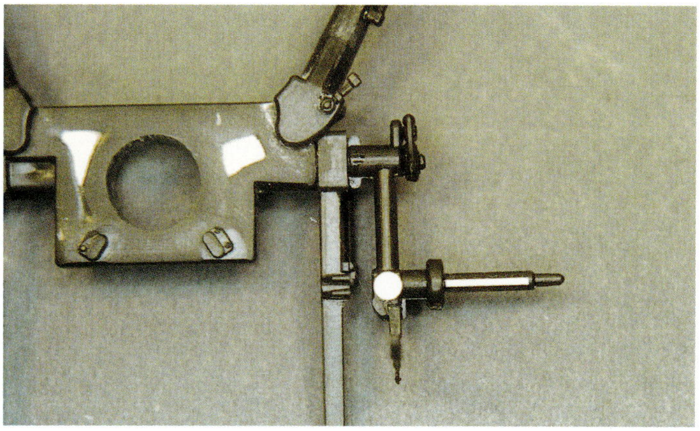

Some of the sink holes and injection marks were in locations that were hard to fix or where the parts were too delicate to work with. In these cases, hide the problems. I used a Waldron punch tool and .005″ sheet stock to make this disk.

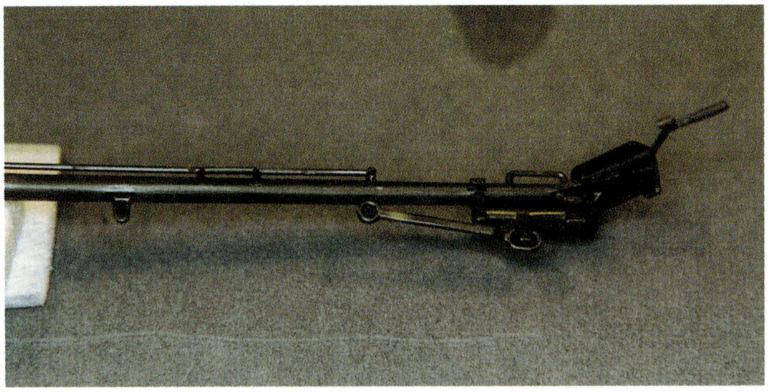

Adding all the small details along the length of the framing of this Italeri 75mm antitank gun was easy, but removing all the mold lines from the parts prior to gluing took some time and care. These small parts are delicate and break very easily.

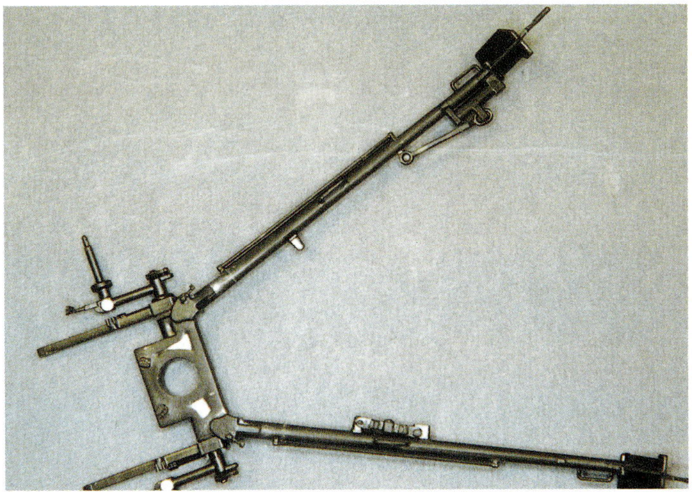

The completed base and framing on this antitank gun is now ready for priming and painting.

The gun shield for this antitank gun is almost a model in itself. Note again the use of punched-out disks to hide flaws.

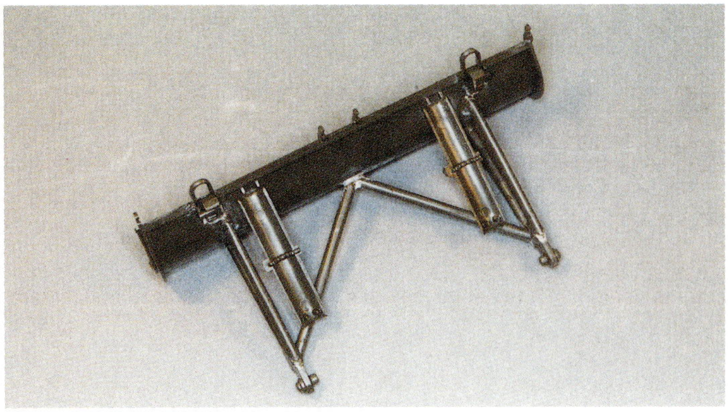

This self-propelled artillery gun shovel had injection marks all over the framing, but here the solution was to scrape the sink marks until they blended into the surrounding plastic.

After completing, test-fitting, and painting all the subassemblies, it's time to start putting it all together. This base, cradle, and slide assembly are ready for the barrel/breech assembly. Note here that the base and cradle were assembled and painted and then the slide assembly was added by popping it into place.

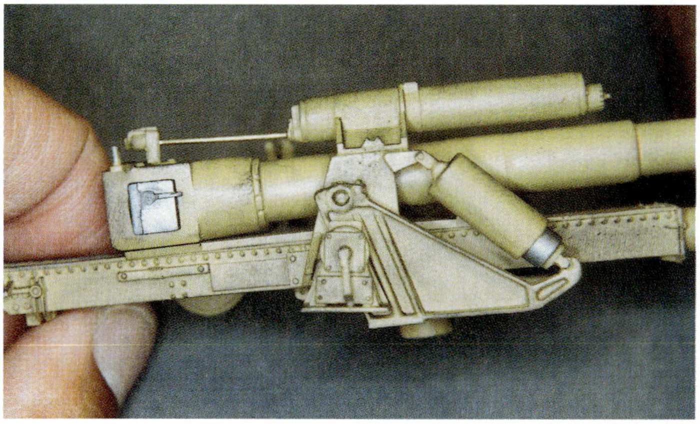

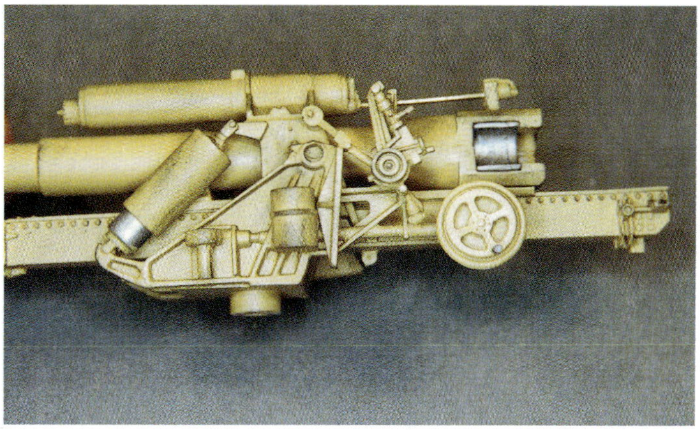

The completed gun assembly on this DML Hummel has been painted and weathered. Note the clean appearance of the assembly. No sink holes, injection marks, or mold lines are visible, and the weathering is very subtle. Model by Bill Teehan

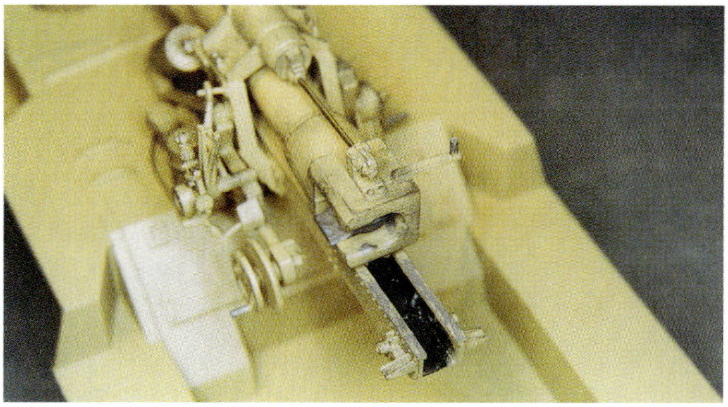

The inside of the slide has a seam that was almost impossible to repair, so its appearance was hidden using a heavy coat of black paint. Note the use of pastel pencil dust to simulate the smoke and soot of gunpowder stains. Model by Bill Teehan

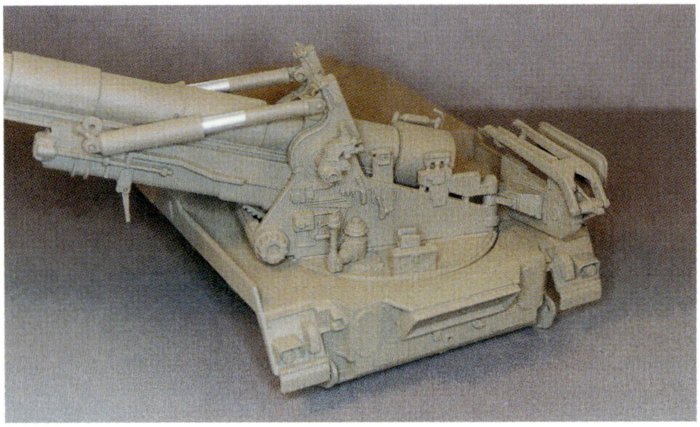

Test-fit the completed gun assembly on this Italeri M-110 on the hull. I checked the clearance on the assembly and found that the shell carriage loader was going to interfere with parts that were to be attached on the right and left sides of the hull. The solution was to install the gun assembly before adding these additional parts. Otherwise, I would have had to break off the shell carriage or the surrounding parts to install the gun assembly.

This DML Hummel gun assembly is getting its final fit check. The next step is to add the sprockets, road wheels, idlers, and tracks and then attach all the fittings on the upper hull. Model by Bill Teehan.

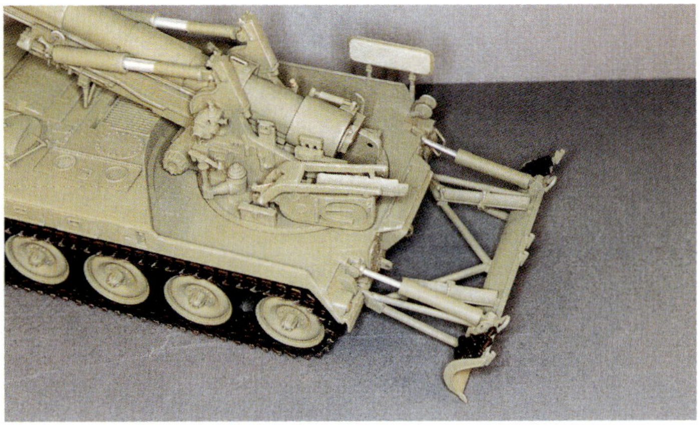

The shovel assembly on this M-110 is now installed and the remaining hull fitting will be the last assembly step. Note the different shades of olive drab. The difference in the colors is very subtle, but your eyes can pick them up, making details and shapes really stand out.

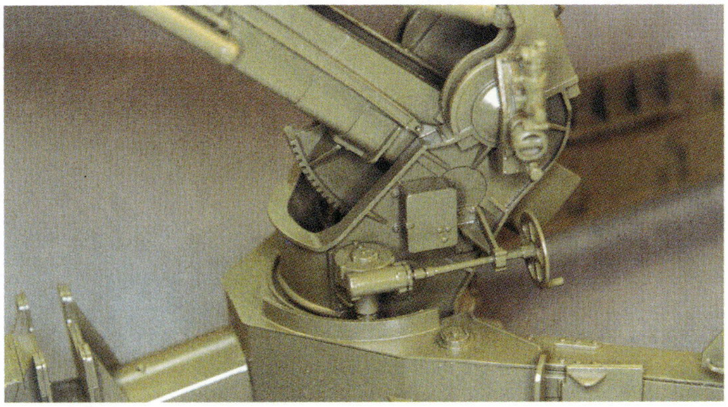

The gear and wheel detailing on the elevation and aiming equipment on this ARV club Long Tom are outstanding. Drybrushing and washes will really make these parts stand out. Note how clean the model appears. It has a gloss coat, because it is ready for decaling. Model by Richard Boutin, Sr.

Paint small parts like these tools separately and then attach them using white glue. Be sure to glue these parts in place prior to glossing the surface, since white glue sticks best to flat painted surfaces. Model by Richard Boutin, Sr.

There is an extensive level of surface detailing and fittings on this Long Tom. Since the framing legs on this model can rotate, it's easy to install the gun assembly onto the slide by pushing the frame legs apart and sliding the gun assembly up into the cradle/slide assembly. Model by Richard Boutin, Sr.

Install the torsion bars on the suspension system of this Long Tom, and the assembly is almost ready for weathering. Model by Richard Boutin, Sr.

The level of detail on this ARV Club suspension system is superb. Every the tiny individual linkage rods are included. Cutting these small, delicate parts from their trees required a lot of patience. Model by Richard Boutin, Sr.

Because the suspension system was made up of many small parts, it was not possible to test-fit it by assembling the main components with masking tape. Luckily, the suspension system fit perfectly.

The Long Tom trailer frame assembly is complete and ready for its tires. The travel lock on this assembly also rotates. Model by Richard Boutin, Sr.

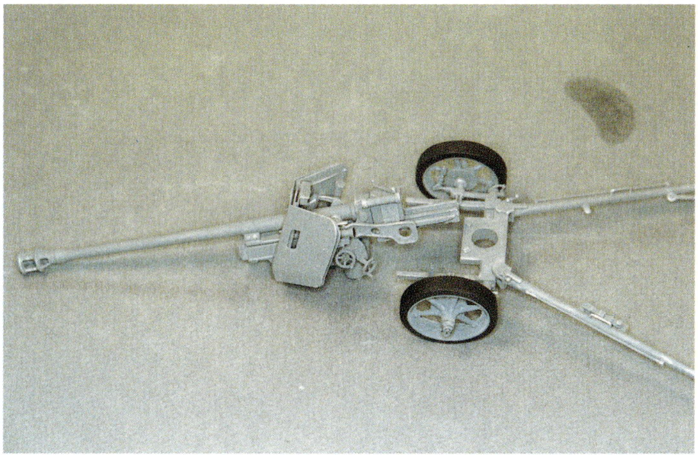

This 75mm antitank gun assembly is a good example of how to deviate from the instructions. The base, frames, and wheels make up one assembly and the gun, shield, slide, and cradle the other. After the two subassemblies are attached, the completed model will get several dilute coats of mud, dirt, and dust.

Note how a dilute coat of Testors umber color in combination with drybrushing brings out all the detail on this Long Tom. Model by Richard Boutin, Sr.

The dilute umber paint settles into the nooks, inner edges, and corners so that these details also stand out. Model by Richard Boutin, Sr.

The breech assembly of this Long Tom is one of the eye-catchers on this model. Note that the ribbed area of the breech is a much lighter color so that it will stand out. Model by Richard Boutin, Sr.

tips

- Don't forget to drybrush all the edges with Testors silver paint.
- Black pencil pastel dust applied with a soft brush onto flat paint works well for gunpowder burns.
- Seal pastel dust with Testors clear Dullcote.

This antitank gun has received its dilute coats of mud, dirt, and dust. Notice that the black rubber tires now appear very dirty, yet you can still tell they are black.

Add the suspension system, complete with tires, to the Long Tom. If it were not for the umber color wash and the drybrushing with silver paint, the little details on this model would not be visible. Model by Richard Boutin, Sr.

Note the subtle drybrushing around the edges of the gun shield and on the bolt heads and hinges.

Here again, weathering and drybrushing help make details stand out. Ink the sighting tube face with a fine-tipped disposable drawing pen.

Tanks and Artillery Gallery

M4 Sherman. 1/35 scale Tamiya kit built by Daniel O'Keefe

Zvezda JS II M. 1/35 scale Italeri kit built by Jim Hudson

Pzkpfw1 Ausf.F. 1/35 scale Alan, Inc. kit built by Jim Hudson

V-100 armored car. 1/35 scale Verlinden Productions kit built by Mike Ashey

M-110 self-propelled gun. 1/35 scale Italeri kit built by Mike Ashey

Jagdpanzer. 1/35 scale Tamiya kit built by Scott Weller

75mm antitank gun. 1/35 scale Italeri kit built by Mike Ashey

Opel Blitz 3(T) Type S. 1/35 scale Heller kit built by Mike Ashey

Right: M-18 Hellcat. 1/35 scale Academy kit built by Mike Ashey

Top and above: M-59 75mm Long Tom. 1/35 scale ARV Club kit built by Richard Boutin, Sr.

King Tiger with Porsche turret. 1/35 scale Tamiya kit built by Daniel O'Keefe

Leopard II. 1/35 scale Italeri kit built by Jim Hudson

MK-IV World War I British female heavy tank. 1/35 scale Emhar kit built by Mike Ashey

F. 4

T-28 Soviet tank. 1/35 scale ICM kit built by Steven O'Connell

Tiger I (Tunisia). 1/35 scale Tamiya kit built by Keith Lehle

British Challenger tank. 1/35 scale Tamiya kit built by Jose T. Rodriguez

Diamond T. 1/35 scale Scale Model Accessories kit built by Owen R. Auer

FV-107 (T) Scimitar. 1/35 scale AFV-Club kit built by Matthew W. Malogorski

T-34-85 Soviet tank. 1/35 scale Tamiya kit built by Jim Hudson

Jagdpanzer Mk-IV/70 Lang. 1/35 scale DML kit built by Bill Teehan

British Mk 1 female tank. 1/35 scale Airmodel resin kit built by Jim Hensley

M3A1 scout car. 1/35 scale Peerless kit built by Bob Skladany

Russian 203mm tracked howitzer. 1/35 scale MB Models kit built by Edward C. Kennedy

Sd. Kfz. 165 Hummel. 1/35 scale DML kit built by Bill Teehan

T-72 tank. 1/35 scale Kirin kit built by Jim Hudson

5 Building and Detailing Military Vehicles

Military vehicles have various levels of detail on their chassis, motors, and crew compartments, and adding small scratchbuilt details can really enhance their appearance. One of the vehicles I chose to build for this book is Heller's 1/35 scale Opel Blitz Type S truck. It has a nicely detailed chassis and a full engine plus a radiator. The engine detail is not complete, and there is no detail inside the engine compartment or firewall, but with bits of Evergreen plastic rod and strip, a few parts from your spare parts bin, and some wire from Detail Master, you can make the engine compartment the centerpiece of the model.

I used two reference books to help scratchbuild the interior parts, and I noticed that the engine wiring in the photos seemed to be a lot thicker than normal. To simulate this, I used Detail Masters 1/24 scale engine wiring for the spark plugs and the other wiring. The distributor was also missing on the engine, so I used Evergreen rod stock to simulate the stem and the distributor. I chose a slightly larger-diameter rod for the distributor so I could drill out the six spark plug wire holes plus the wire that goes to the ignition coil. I also added fuel lines, carb links, and battery wiring, as well as the wiring coming from the backs of the dashboard dials.

The kit's radiator was nicely done, but the filler tube was in the wrong place, so I made a new one. The intake and return hoses for the radiator were not included, but it was easy to scratchbuild them. I used small-diameter brass wire for the radiator pipe filler as well as the intake and return hoses. Your Waldron punch tool comes in handy here for making disks for the back side of the dashboard dials, the oil dipstick tube and cap, and radiator cap. I simulated the dashboard dials with decals and scraped off the molded-on gas, clutch, and brake pedals and fashioned new ones with scraps of Evergreen strip and rod.

Building up this model is also a good example of how you have to plan your work and deviate from the instructions. Cutting out the driver's door and the left front hood so the engine area would be exposed was not difficult, but assembling the cab with these parts cut out was a bit tedious. I had to position the parts, tape them in place, and make several adjustments before I could glue the cab sections. I also had to glue the cab together in such a way as to leave enough room to work in these areas prior to final assembly. Thinking it through and working out the assembly before I did any gluing made for a very successful experience.

The other kit I built up was a Verlinden armored vehicle resin kit. Resin kits can be easy to build because all the parts are usually cast as complete pieces. While there are no seams or cast lines to take care of, you will have casting blocks to remove and resin casting imperfections to repair. Good resin manufacturers don't have a lot of imperfections, but occasionally you get a bubble void. Sometimes you can get lots of bubble voids in places you won't see once the vehicle is assembled. When this happens it is usually an unavoidable result of the casting process.

Resin tips. Working with resin is easy as long as you follow a few simple guidelines. First, be very careful when removing resin pour plugs from parts. Resin is easy to cut and sand, so don't overdo it. I recommend cutting the plugs down to as small an area as possible using a razor saw, then sanding off the remaining resin. The excess resin on parts can be easily sanded off by running the part across a stationary piece of sandpaper in a figure-eight motion. Rotate the part so you don't sand off too much from one side. Take your time and check your work frequently. Once you have removed the excess resin, scrub the parts with a soft toothbrush and mild soap and warm water to remove mold release agents and resin dust.

Wear a dust mask when sanding resin—the resin dust particles should not be inhaled. One way to reduce dust is to always wet-sand resin (always use waterproof sandpaper for this). I usually use thumb tacks to hold the sandpaper still and flat, and I have a special length of pine, which is perfectly flat, that I pin it to. Then I dip the part into water and begin sanding, frequently dipping the part to remove the resin slurry and to add more water. Another benefit to wet-sanding is that it keeps the sandpaper from becoming clogged with resin, so it will last longer.

Bubble voids are easy to fill with super glue, putty, or Evergreen strip stock. For voids on flat or large curved surfaces, use thick gel super glue or resin. Testors putty also is great for this. For small voids use automotive crack filler or a thin super glue applied with a small-diameter wire applicator. Sometimes the voids are so small that the surface tension of the super glue won't allow it to seep into the tiny area. In these cases, enlarge the hole slightly with a drill bit if you are using super glue, or else use automotive crack filler. I have found myself using automotive crack filler more and more these days to deal with small voids because this stuff is easy to use, dries fast, wet-sands easily, and blends perfectly with resin. Super glue accelerator won't affect resin, so you can use it to speed up the drying of super glue.

You can also use Evergreen strip stock to fix small shapes and correct problems in corners, on edges, or on the rims of circular shapes. Use a strip size that fits into the hole, dip the tip of the plastic in a puddle of super glue, and insert the tip of the strip into the hole. When the glue is dry, cut the plastic and trim and sand to shape.

Once you have fixed any problems, give the parts a final cleaning with soap and water and apply a coat of primer. The primer will act as a final check for any voids or bubbles you may have missed. They are easy to miss, especially with light-colored resin parts, because the light color will tend to make your eyes "snow blind." Once the parts are completed, they are ready for final painting and assembly.

Metal tips. To close out this chapter, let's talk about working with brass photoetching and white-metal and brass castings. Always cut photoetching with a sharp blade and cut the parts on a hard surface such as a glass or Plexiglas plate. Some manufacturers suggest in their instruction sheets that you can cut off parts with metal scissors. I recommend against this—it is too easy to damage parts or to cut into adjacent parts. Generally, it's easier to cut a photoetched part off its tree by leaving a little of the stub attached to the part. You can easily remove the remaining stub from the part by using a number 11 X-acto blade and then smoothing the edge with a Flex-I-File sanding stick. Generally, photoetched sheets are between five thousandths (.005) and seven thousandths (.007) of an inch thick, but I have also worked with brass sheet as thin as three thousandths (.003) of an inch. With photoetched sheets being made thinner and thinner, you have to be careful not to distort the parts as you cut them off the trees and clean them up. An important point here about cutting photoetching—wear your glasses or use a pair of safety glasses. Small photoetched parts, especially the stubs, have a bad habit of becoming projectiles when they are cut, so protect your eyes!

Once the part is cut off, clean the surfaces of the part by lightly running it across a stationary piece of 600-grit sandpaper taped to your workbench. Some modelers prefer to clean the entire detail sheet at once by laying it flat and carefully running the sandpaper over it. While this may be quicker, you also stand a good chance of damaging individual parts, so I recommend doing them one at a time. Don't slide the sandpaper back and forth, and only a few passes at the surface will be needed to clean it up—when it's shiny, it's clean. Cleaning up photoetched surfaces is a must if you are gluing sections together.

To get curves in photoetching, bend the part around a wood dowel or other round object that has a slightly smaller diameter than you need. The photoetching will spring back a little, so the smaller diameter is important. You can get sharp bends and angles by using a set of flat needle-nose pliers, or by using two single-edged razor blades.

Super glue is an excellent bonding agent for both brass and photoetching, and when strength is not an issue, white glue also works well. Using white glue will allow you some flexibility in positioning pieces together correctly, but be sure to prime the parts prior to gluing, as the white glue will stick better to flat paint than to bare metal. White glue is also an excellent filler for cracks and voids when you're using photoetching. Be sure to prime photoetching before you give it a finishing coat of paint.

Brass and white-metal castings are easy to work with. They can be scraped, sanded, and shaped just like plastic. These metal castings can have mold release agents on them, so give them a good cleaning with an enamel-based paint thinner. Like injection-molded parts, white-metal parts typically have very small seam lines, although they can be difficult to see because the metal is shiny. Scrape and sand the seam lines just as if they were plastic. Brass castings can also have seam lines, but the metal is much harder, so scraping may not remove them entirely. Use sanding sticks or even a small fine-tooth file. When you are satisfied with your work, give the parts a coat of primer. This will highlight any seam lines you may have missed. Brass and white-metal parts are soft metals, and sometimes they may be bent when you get them. To bend them back into shape, carefully work out kinks with a set of flat-faced needle-nose pliers.

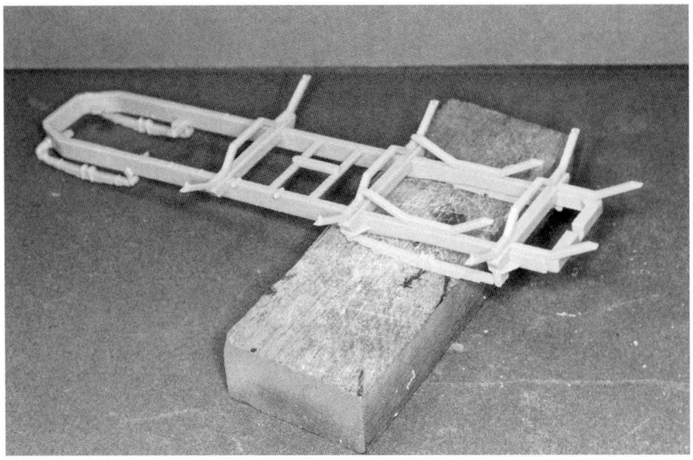

Military vehicles can be a lot of fun to build. The frame and leaf springs on Heller's Opel Blitz have been assembled, and the next step will be building up and detailing the cab, engine, and engine compartment.

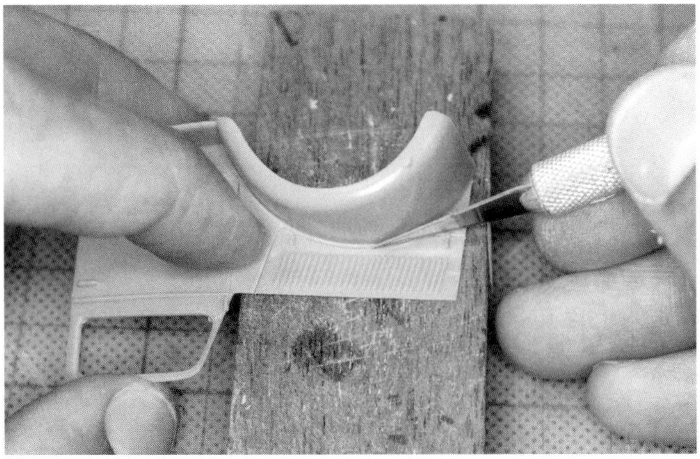

If you plan to detail the engine compartment, you'll have to cut out the hood. Since the scribe line that outlines the side hood is distinct and deep, use the tip of a number 11 X-acto blade to cut out this part.

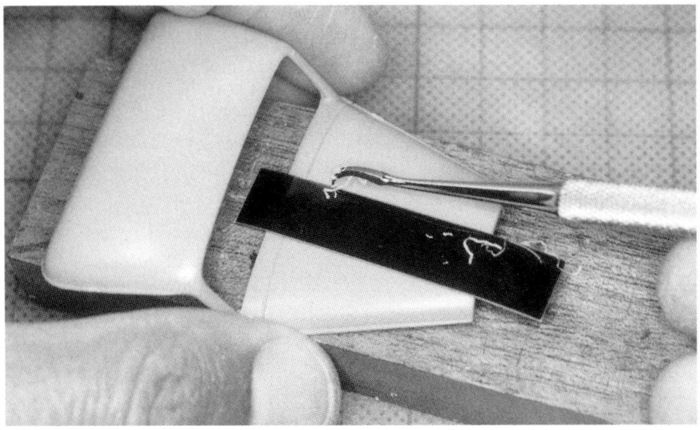

Since the hood top does not have a distinct engraved line, use Bare Metal Foils plastic scriber to cut out this part. Labeling tape acts as a guide for the scriber.

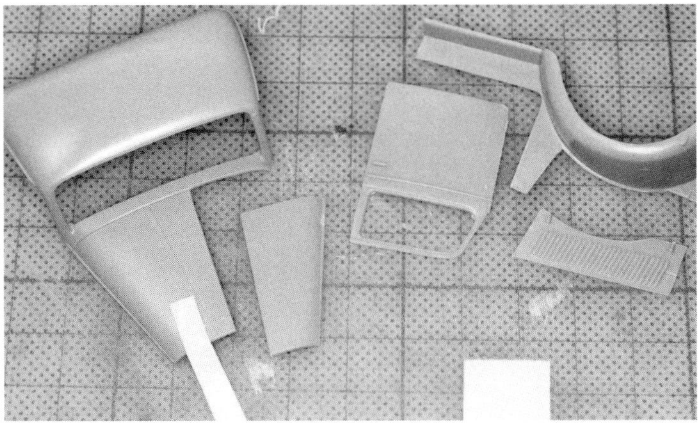

Cut out the hood parts as well as the door, and clean the edges.

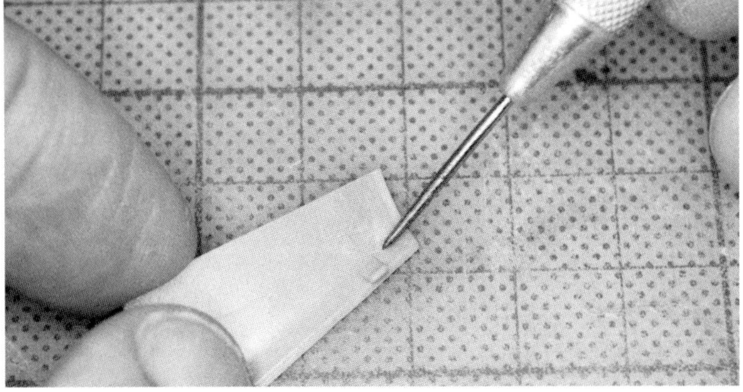

When you cut out parts to display interior areas, don't forget to check the undersides for positioning tabs. Remove a tab like this with a stencil knife.

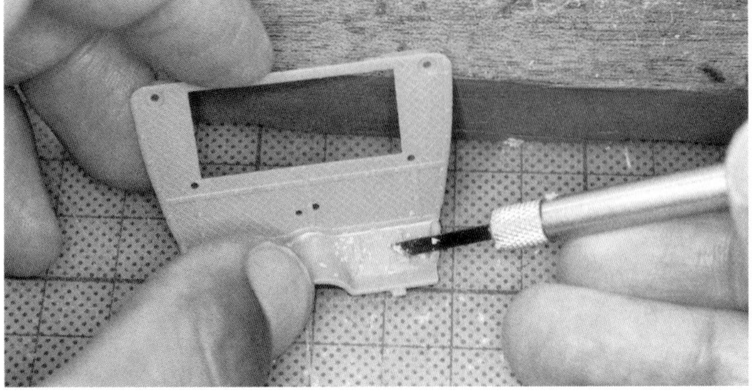

Carefully remove the molded-on gas, clutch, and brake pedals from the floorboard using a stencil knife. You will scratchbuild new ones.

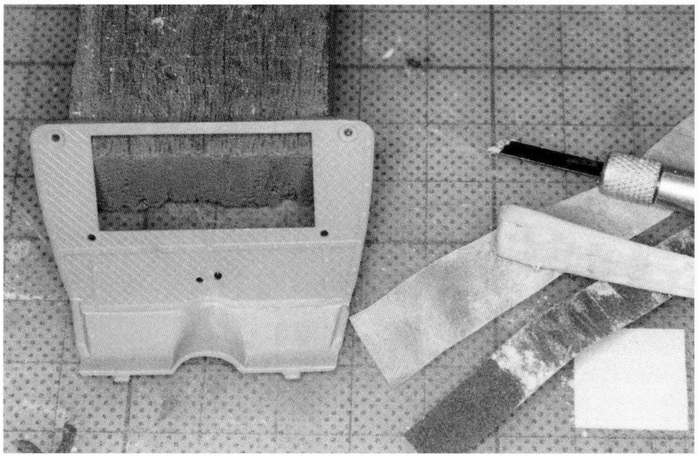

Sand smooth the floorboard area with various grades of sandpaper.

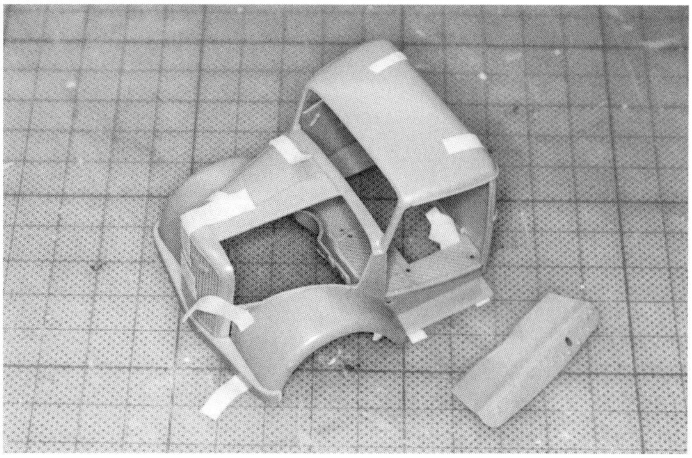

The cab must be assembled in such a way that you will be able to access the interior of both the cab and the engine compartment to add details. This is another good example of deviating from the instructions and creating your own assembly steps.

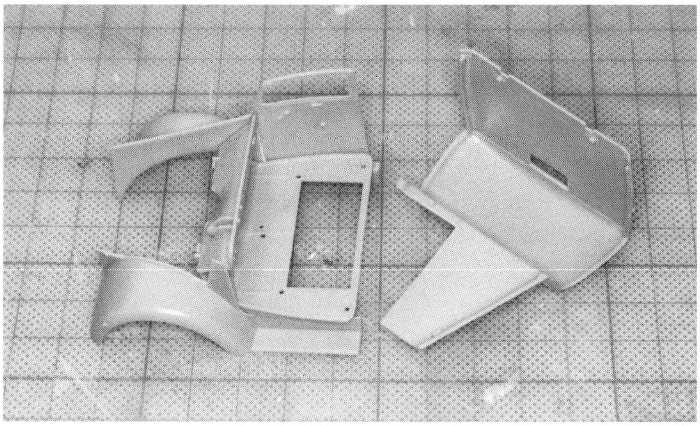

Glue the cab together into two subassemblies.

Once the engine block is assembled, temporarily place both the engine and the forward cab area on the chassis so that you can accurately position the steering column.

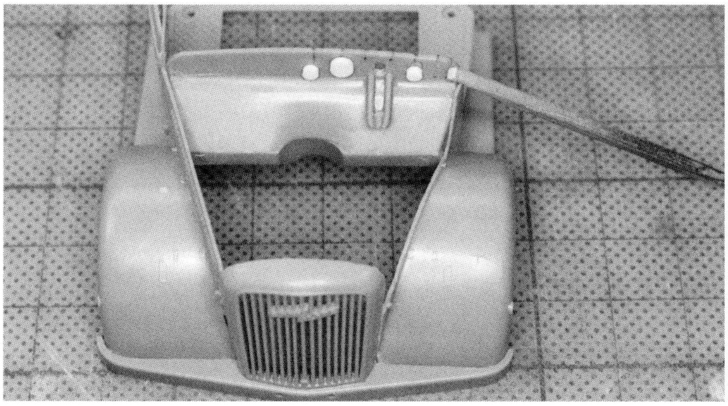

Add plastic disks punched out with a Waldron punch tool to represent the back side of the dashboard instruments.

Use a NorthWest Short Line chopper to make the battery shelf parts. The chopper is a great tool for making identical multiple parts.

Use a NorthWest Short Line True Sander to true up the sides on this small part, which will become the fuse box.

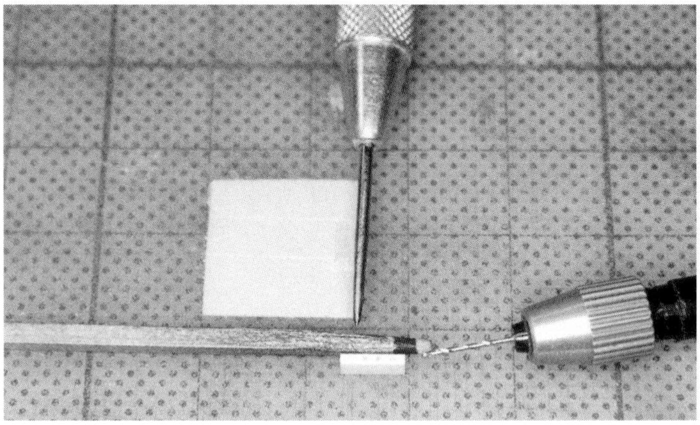

Use a tiny drill bit to drill holes on the sides of the plastic strip for the wires. Once you glue this part into place and insert the wires into the holes, it will look just like a fuse box.

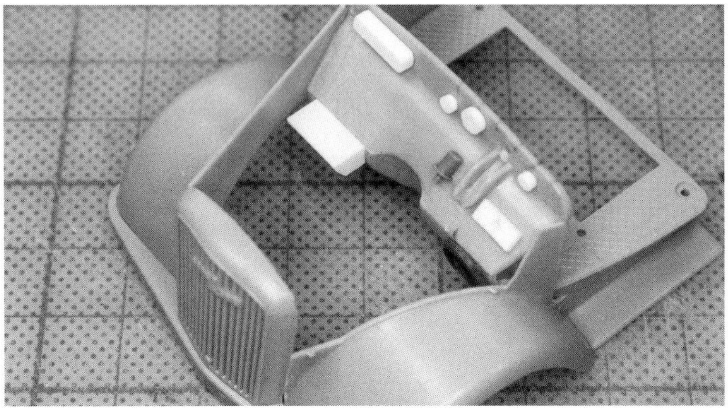

Here the dashboard and fire wall have received all their scratchbuilt components, including the back side of the glove box.

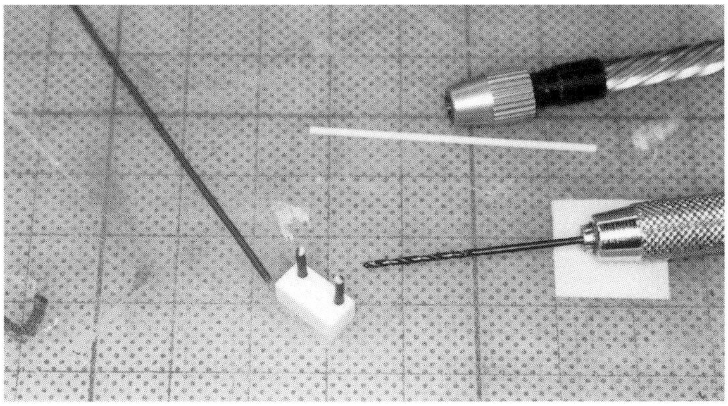

Glue together two pieces of thick Evergreen strip to create the right thickness for this future battery. Cut down the length using a razor saw and miter box, then true up the sides with a True Sander. Install the battery posts as in this photo. The next step is to trim them to the proper height.

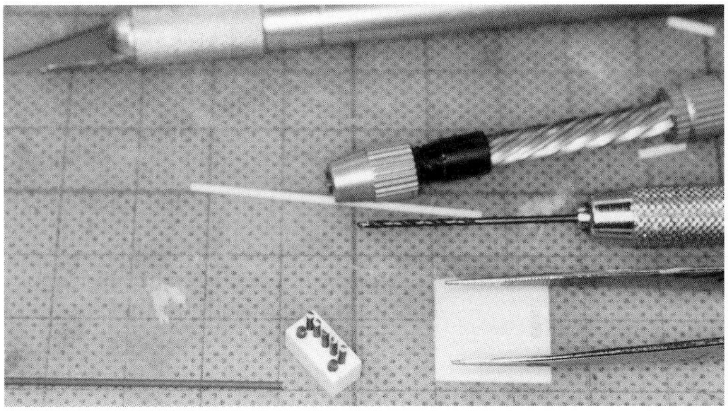

Add the battery cell caps, and trim them to size. Use a sanding stick to carefully sand down the stubs to the proper height.

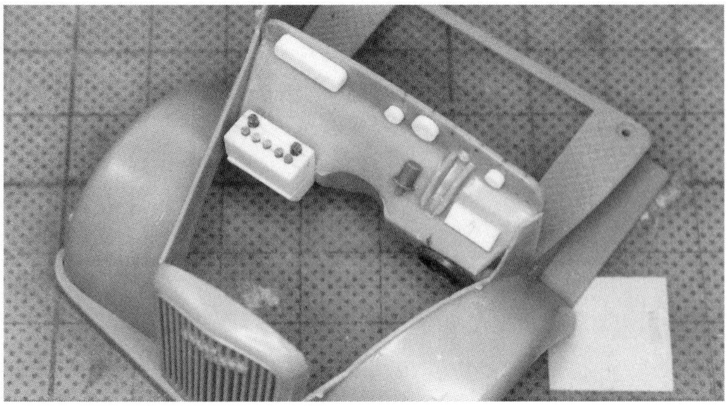

Here the battery gets one last fit check on its shelf. Temporarily tape the rear half of the cab into place to ensure that the battery doesn't interfere with the hood.

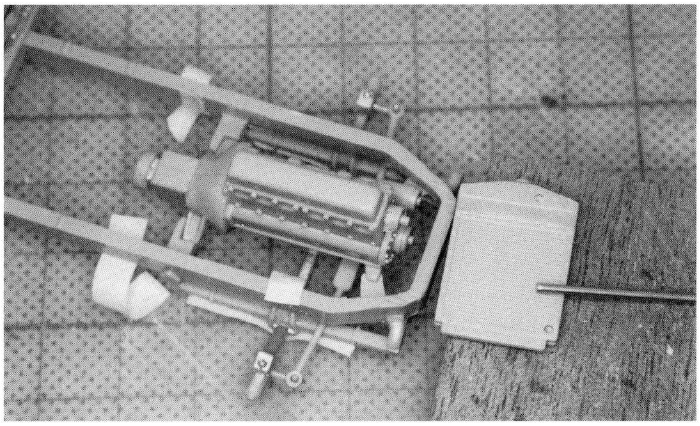

The model did not come with intake and return radiator hoses, but they are easy to add using brass wire.

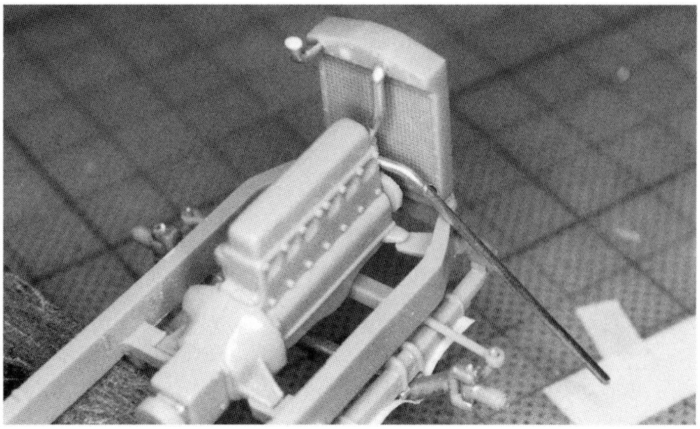

After giving the lower hose a few more bends, cut it to size. Form-fit both hoses using a trial-and-error process. Also notice that I moved the filler tube. This is because the molded-on filler tube interfered with the hood.

The brass radiator hoses are complete. Once you paint them, they add yet another dimension of detail to this engine compartment.

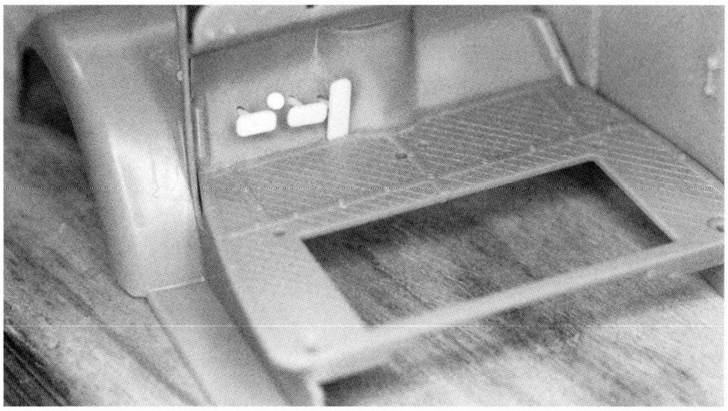

Make the clutch, brake, and gas pedals from scraps of Evergreen strip and rod stock. They look much better than the original molded-on pedals looked.

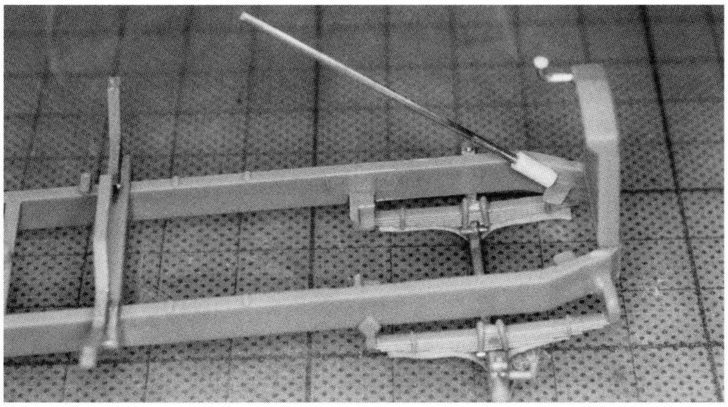

Positioned the steering linkage tube and glue it into place.

Attach the manifold to the exhaust pipe using a small length of plastic tubing. The tubing simulates couplings that you sometimes find on real vehicles.

The drive shaft on this kit was too short, but a short length of Evergreen rod fixed the problem.

Complete and paint the frame and add the tires. Once the engine is installed, it's going to look like a real vehicle.

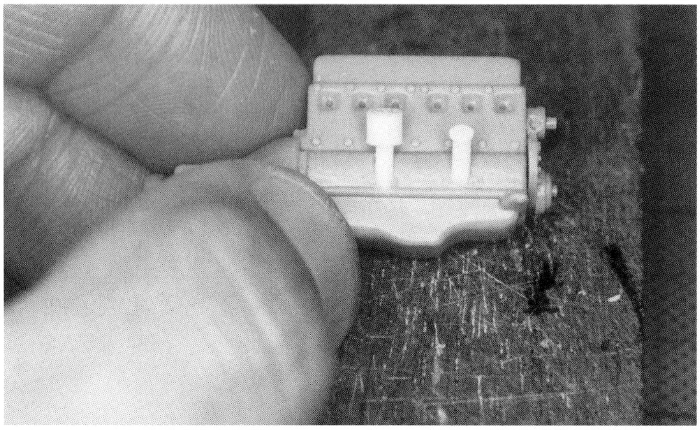

The kit's engine did not have either a distributor or an oil filler tube. Reference pictures showed both of these parts to be on the right side of the engine. Make the distributor from two different diameters of Evergreen rod and the filler cap with a Waldron punch tool.

If you plan to wire the engine, drill six spark plug wire holes as well as the ignition coil wire hole in the distributor cap.

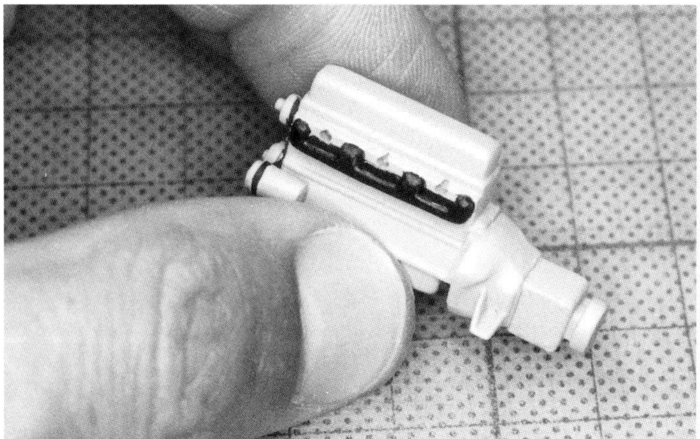

To simulate the rust/burnt iron appearance that you typically find on manifolds, use Testors Metalizer burnt iron.

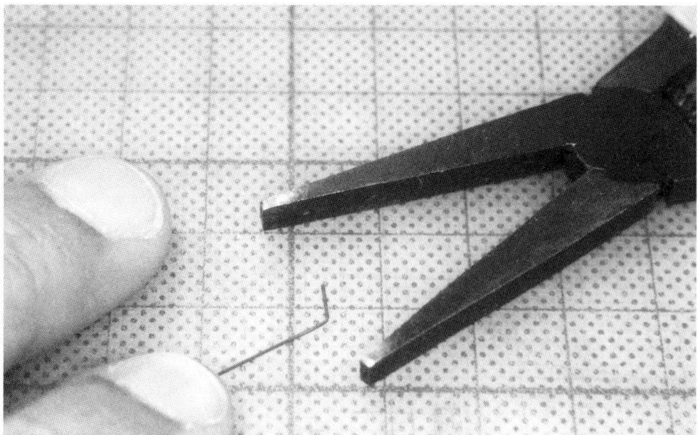

When wiring the engine, use flat-tip needle-nose pliers to make the necessary bends in all your wiring.

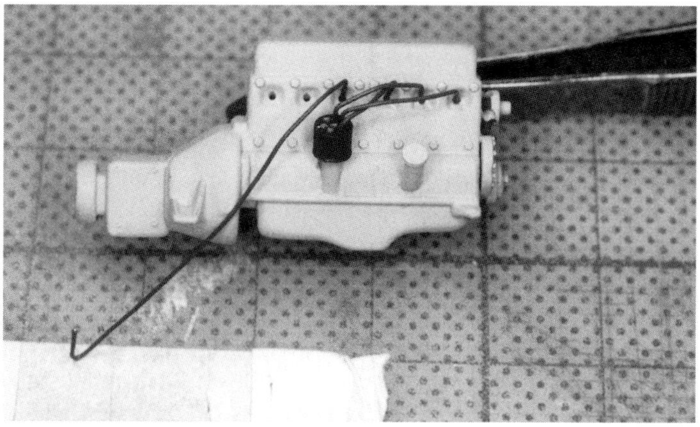

The spark plug wires are progressing. Form-fit each individual wire into place.

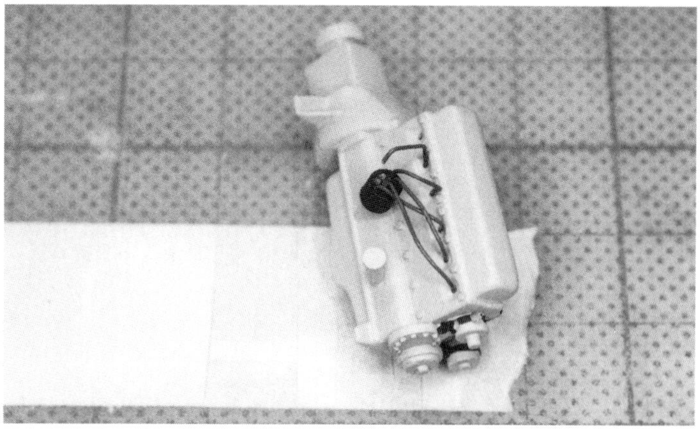

Wiring is just about complete. Note the bend on the tip of the wire that will be inserted into the top of the distributor cap.

Install the engine. The ignition coil wire will hang free until the cab is attached to the body.

Install the carburetor along with the linkage and fuel line, as well as the radiator and its hoses.

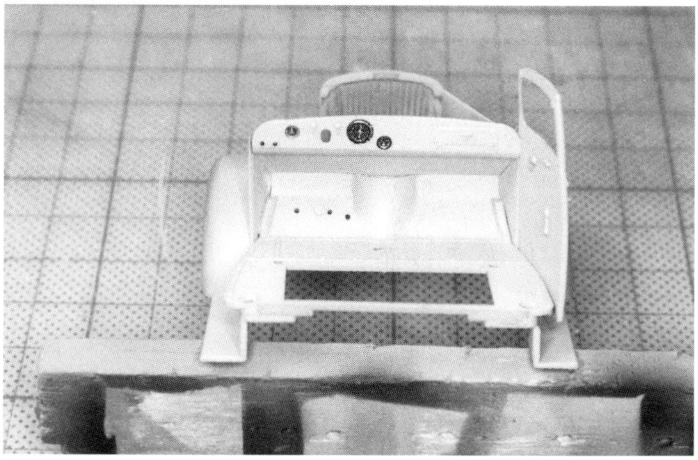

To simulate the dashboard dial faces, punch out decals and attach them to the dashboard.

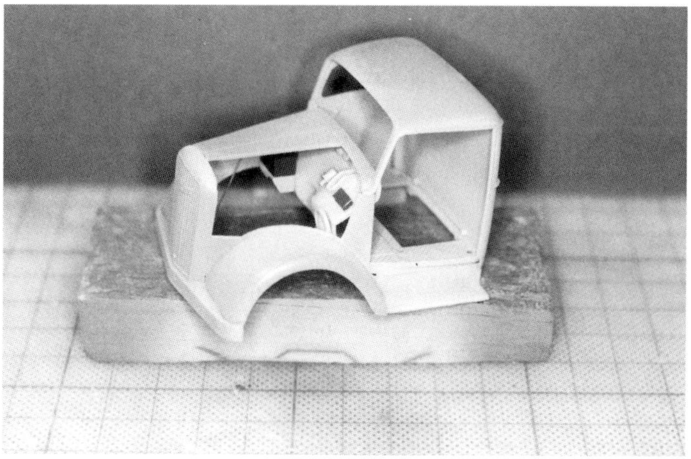

Here is the assembled cab. The next steps are masking and painting.

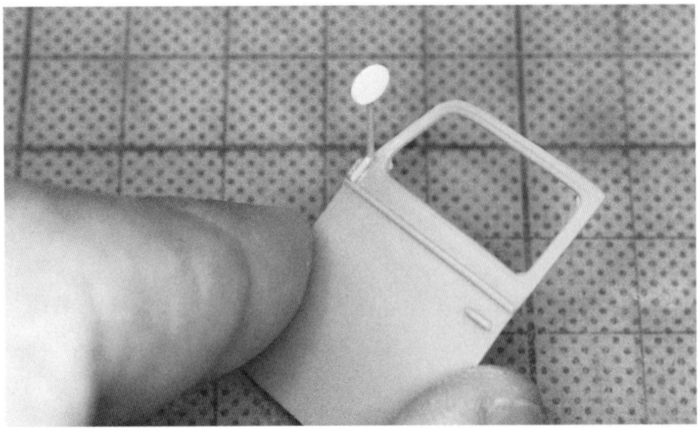

Replace the kit-supplied rear-view mirror with a slightly larger one, courtesy of your Waldron punch tool.

Now install the cab on the frame. This model is really starting to take shape.

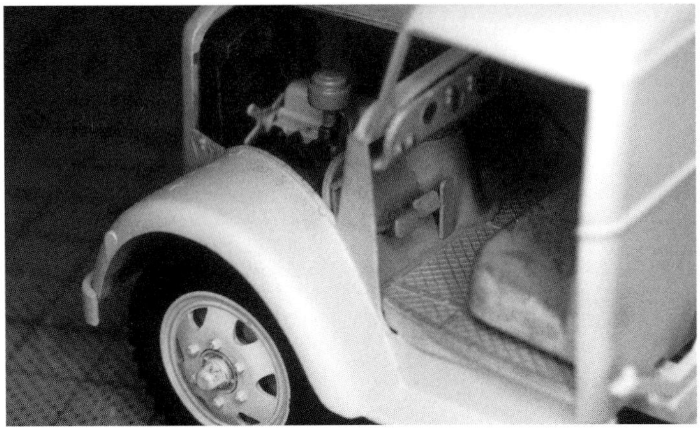

The scratchbuilt clutch, pedal, and brakes add yet another level of detail to the interior cab of this model.

Complete the engine compartment and position the hood. Opening up the hood to display the engine compartment of this model makes it a real eye-catcher.

Attach the canvas top for the truck bed as well as the remaining detail parts. Use dilute coats of dirt- and mud-colored paint to weather this model.

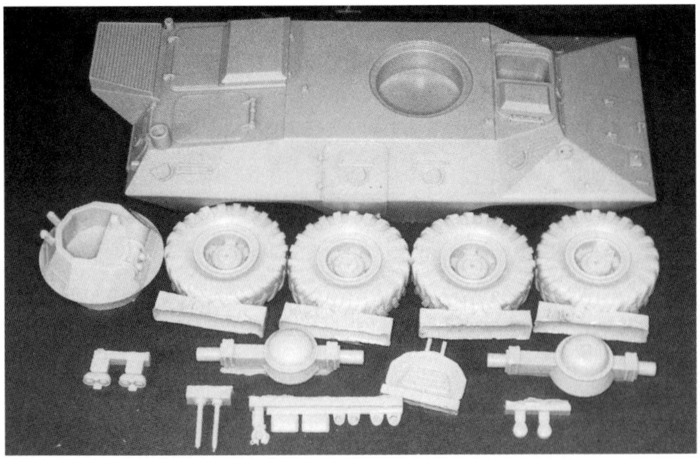

Resin kits usually have complete cast parts that need to be cleaned up and painted. The body on this Verlinden armored car is one huge hunk of resin.

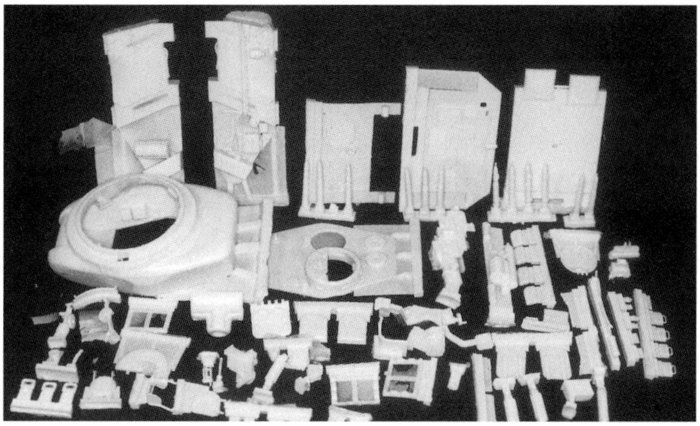

Resin kits can also contain a lot of parts. The major advantage of resin kits is that you don't have to deal with mold or seam lines.

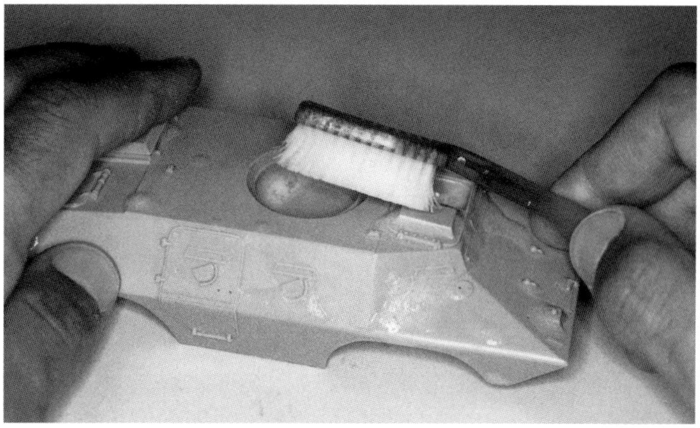

Clean resin parts with warm water and soap to remove all mold release residue.

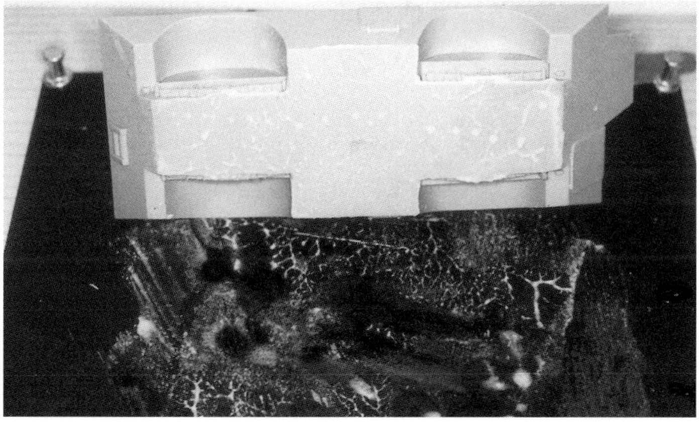

Large, shallow casting blocks on large resin pieces are easy to remove by running the parts across a stationary piece of sandpaper. Always wet-sand resin; then you don't have to worry about inhaling the resin dust.

Use a razor saw to remove the pour blocks from these tires.

To remove the remaining resin from the pour block, run the tires across a stationary piece of sandpaper. Since tread on the tires will be missing because of the pour blocks, these locations will become the place where the tires contact the road.

Use a sanding stick to clean up the sides of the tire where the pour block was located.

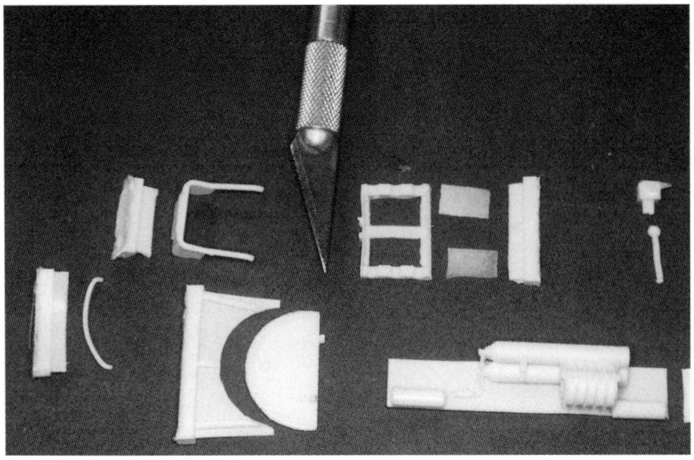

Some resin kit manufacturers engineer the pour blocks on their parts so that they can be easily removed with a sharp number 11 X-acto blade. In some cases, they can be snapped off.

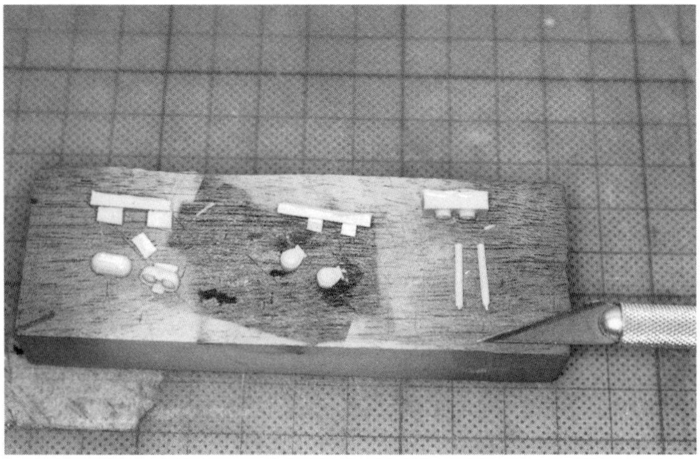

If the pour block attachment points on the parts are thick, as the ones in the photo were, you'll have to remove the parts carefully, leaving excess resin to trim and sand away.

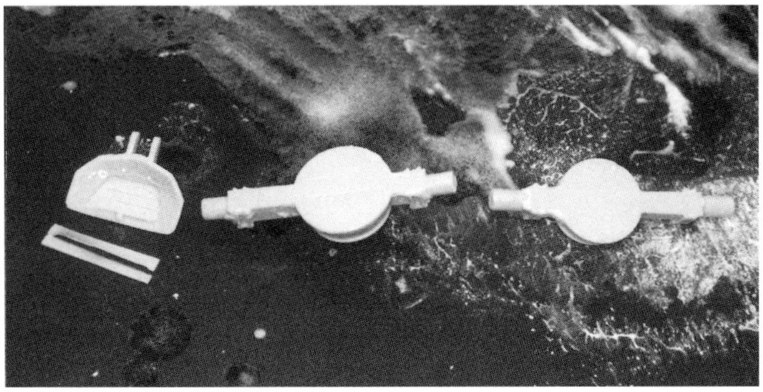

Sometimes you can remove this excess resin by running the parts across a stationary piece of sandpaper. This also works well for flattening out gluing surfaces.

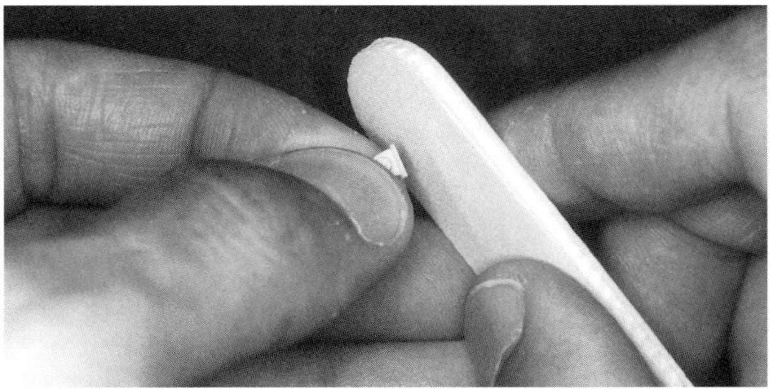

Use a sanding stick to remove the excess resin and to also reshape the part.

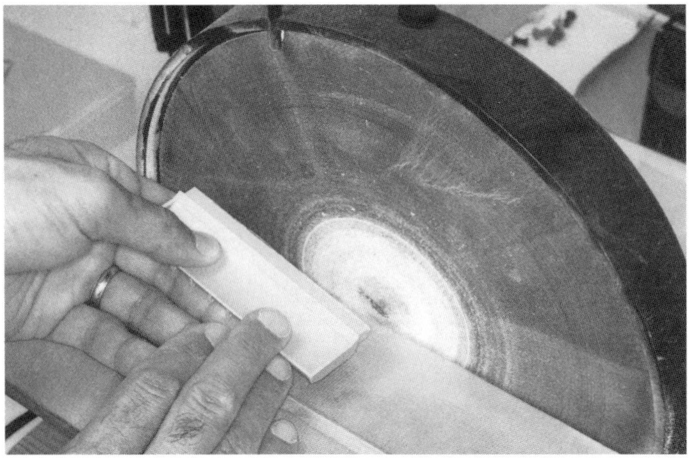

Sometimes you can't cut down pour blocks with a razor saw, and it's too time-consuming to sand them off by hand. In these cases, use a sanding disk.

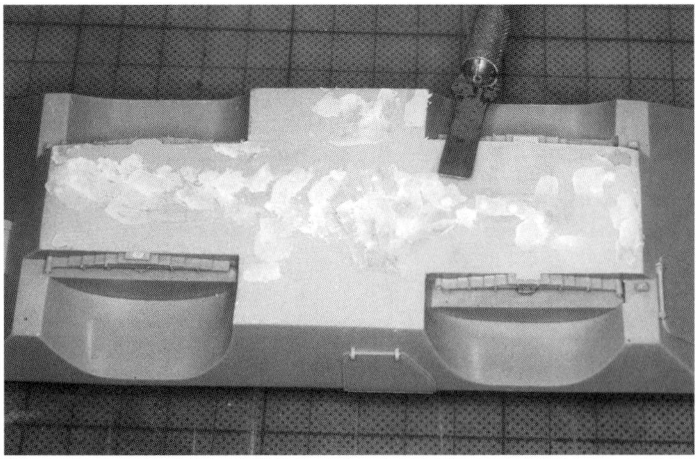

There were bubble voids on the bottom of this Verlinden armored car. Although they would not be visible, I filled them with Testors putty.

To repair bubble voids in small locations, Evergreen rod and strip stock works great as a filler. Dip the tip of the plastic into a puddle of super glue, insert it into the hole, and then trim and sand to shape after it has dried.

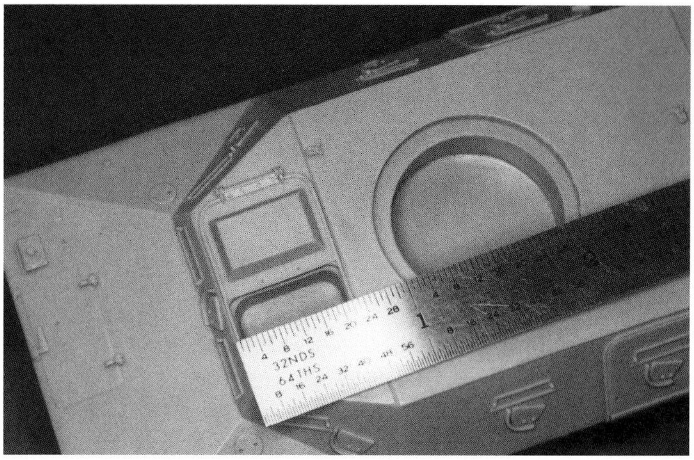

Since I damaged the kit-supplied hatch, I had to make a new one. A machinist's ruler is great for precise measuring.

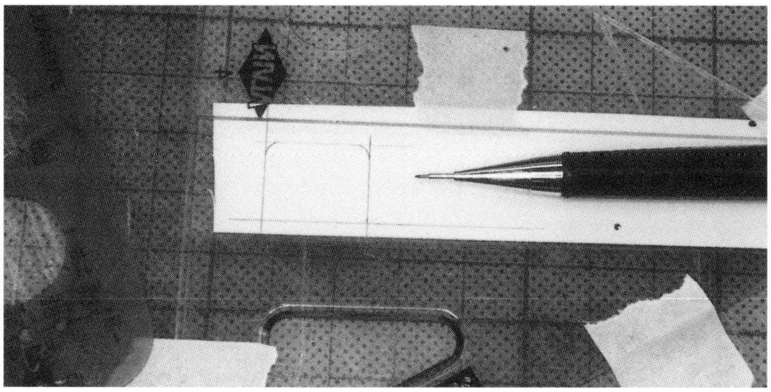

I used small triangles to measure and draw the new hatch cover.

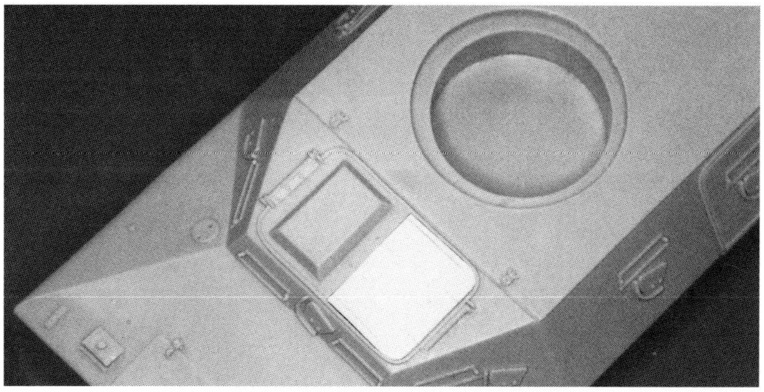

I cut out the new hatch cover and form-fitted it into place.

I could have done the angles on the hatch box using a True Sander, but it was quicker and easier to use my trusty sanding disk. If you want to do any serious scratchbuilding, I recommend that you invest in a variable-speed disk sander with an adjustable table. Both Micro Mark and Model Expo sell these high-quality machines.

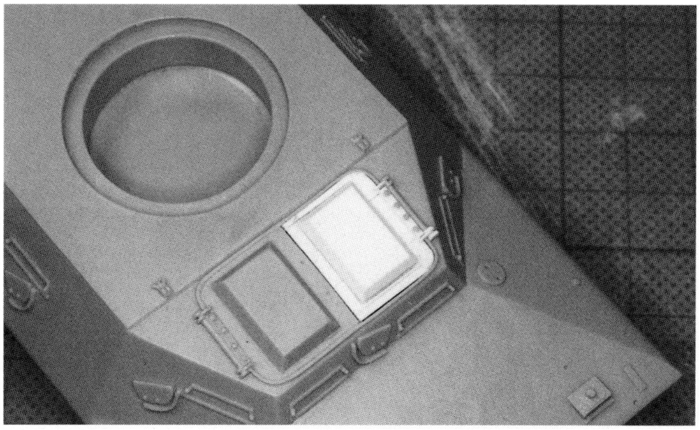

The hatch is now complete and is getting its final fit check. Make the hinges by cutting up disks punched out with a Waldron punch tool. The bolt head details are Grandt Line bolt heads.

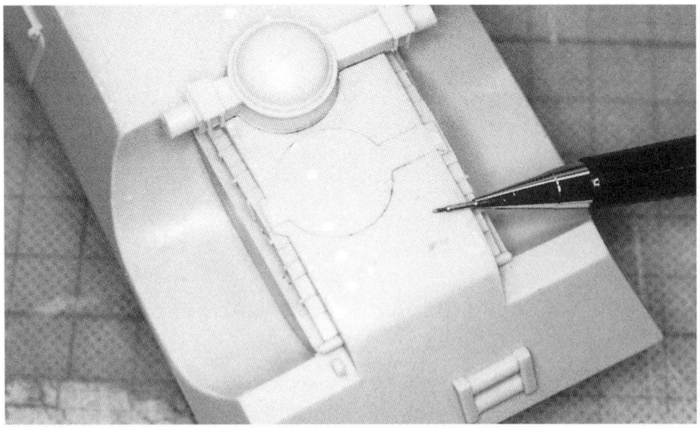

Since there are no location pins on resin kits, position the part and then outline it with a soft lead pencil. When you are ready to glue the parts in place, you can use the outline as a positioning guide.

Install small, delicate resin parts like this lift ring first and then drill out its center. Typically, the resin in the areas that must be drilled or cut out is very, very thin.

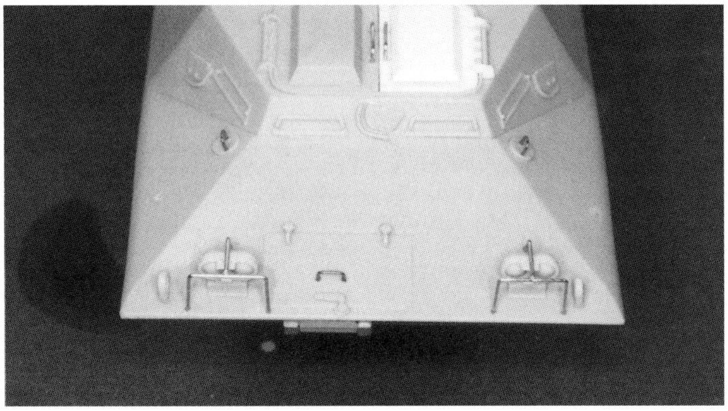

Use stiff brass and soft copper wire to make the brush guards, lift handles, and lifting rings on this model.

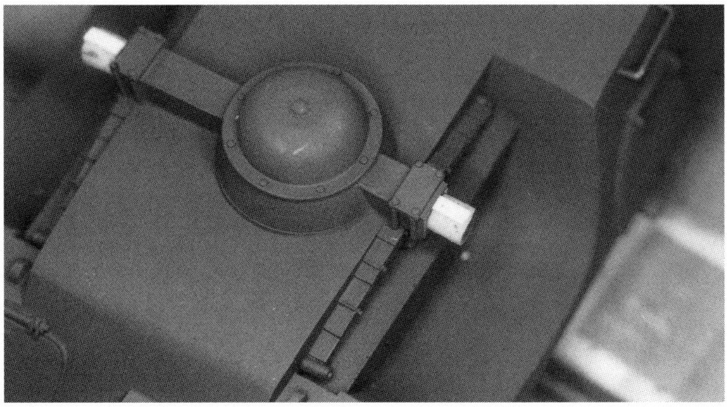

The axles on this Verlinden kit have a smaller diameter than the holes of the tire rims. The simple solution is to add some small strips of Evergreen strip stock to the axles.

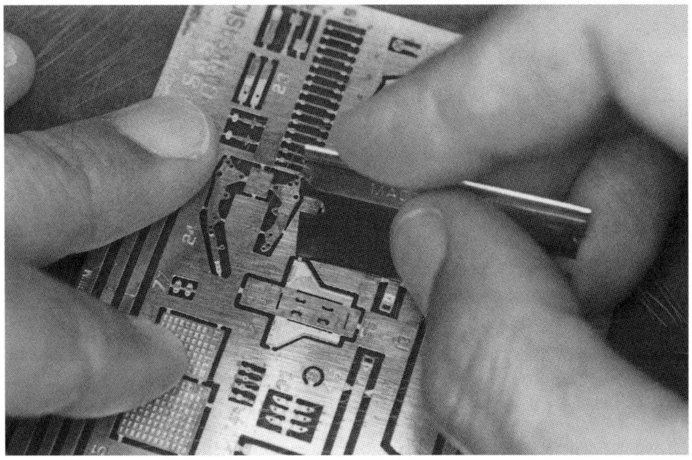

When cutting out brass photoetching, lay the sheet on a hard surface like Plexiglas or a glass plate and use a single-edged razor blade to cut out the parts. Wear eye protection!

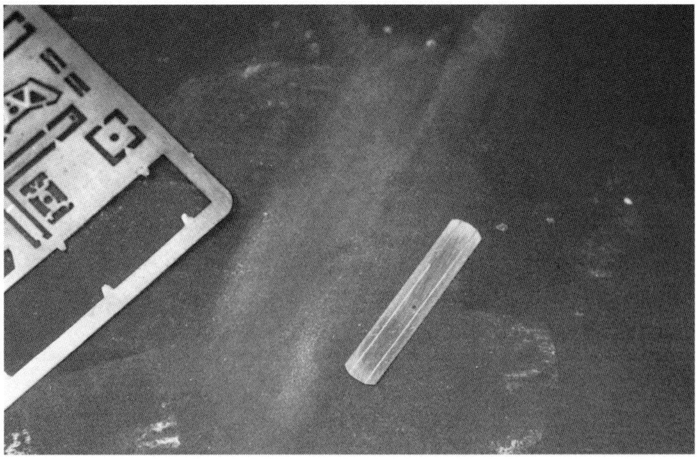

After you cut them out, clean photoetched parts to remove any chemical residue. Run the parts across a stationary piece of fine-grit sandpaper to remove tarnish and chemical residue.

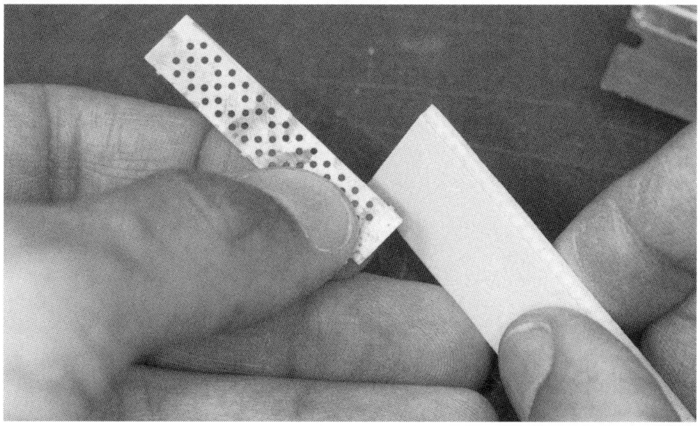

Sanding sticks work well to sand off the stubs that attach the photoetched parts to the trees.

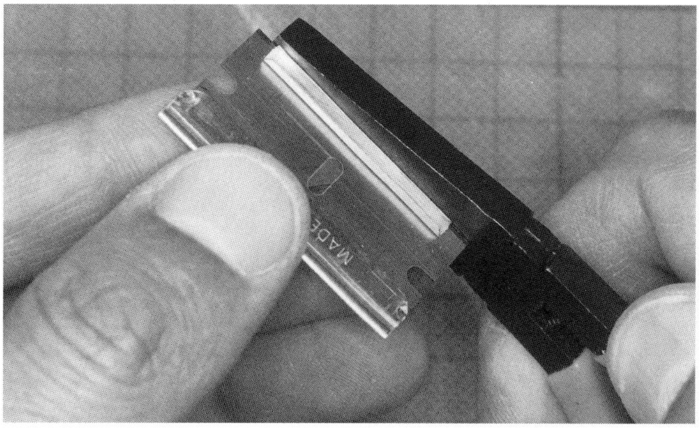

Bend photoetched parts using flat-nosed needle-nosed pliers.

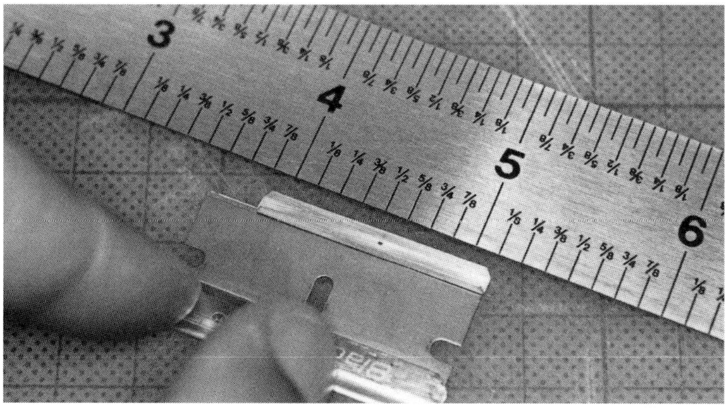

If you need a long bend in a photoetched part, another method is to use a steel ruler and a single-edged razor blade.

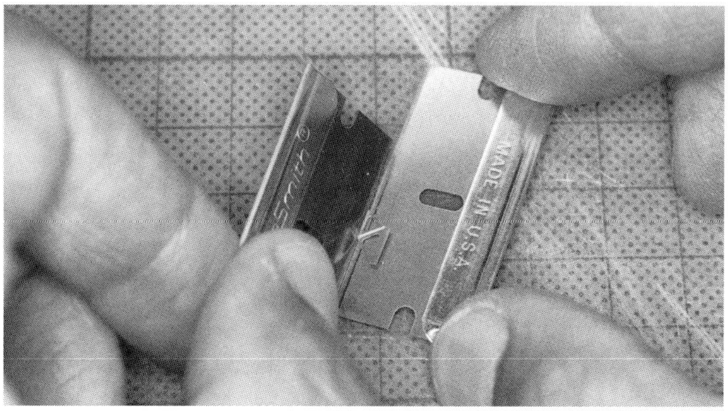

For little bends, use two single-edged razor blades. Position one razor blade at the bend location, slip the other razor blade under the part, and simply rotate upwards.

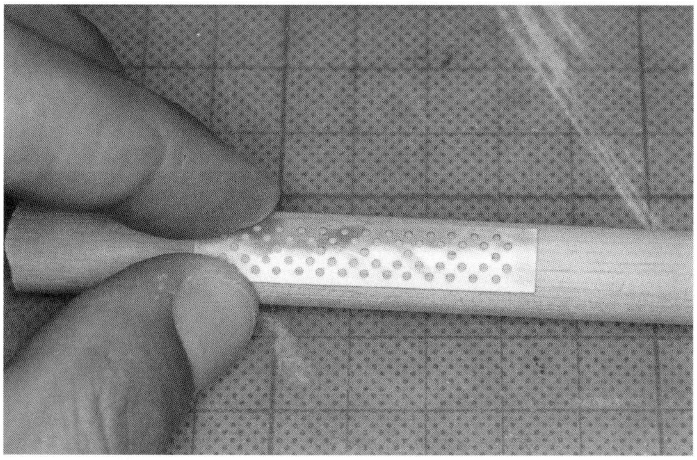

Wooden dowels are great for making curves in photoetching. Use a slightly smaller-diameter dowel, as the photoetching will have a tendency to spring back.

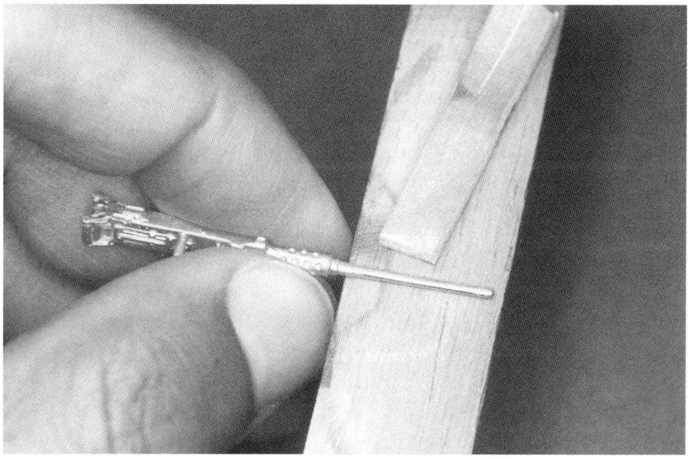

Sand the seam lines along brass castings—the brass casting will not respond well to scraping with an X-acto blade.

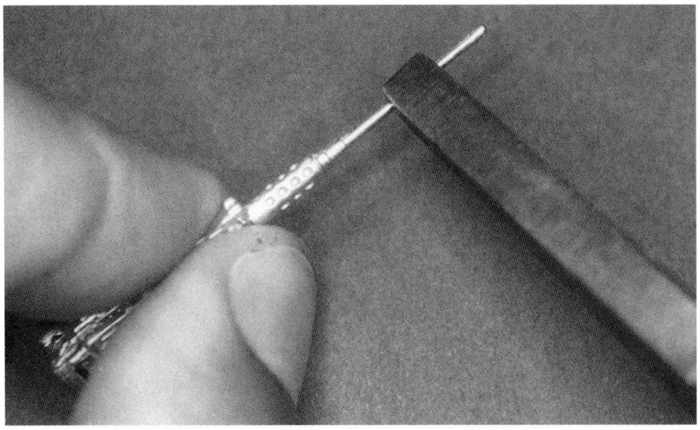

Since thick brass is fairly stiff, it takes a little work to straighten out cast-brass gun barrels. But because the brass is stiff, you will not typically find brass barrel lengths that are wavy in appearance, as white-metal ones can be.

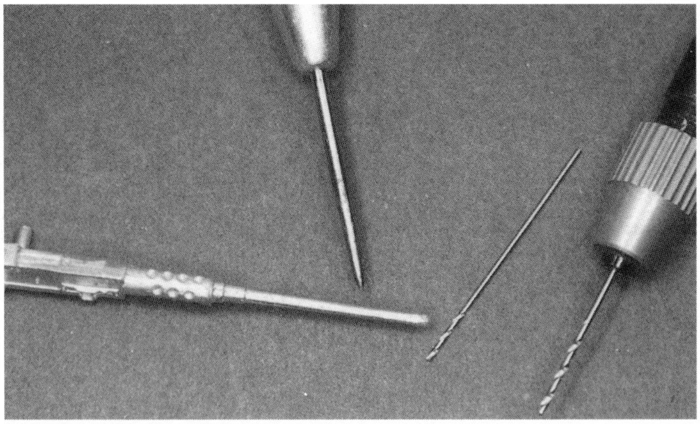

You can also drill out cast-brass barrels. You just have to be patient.

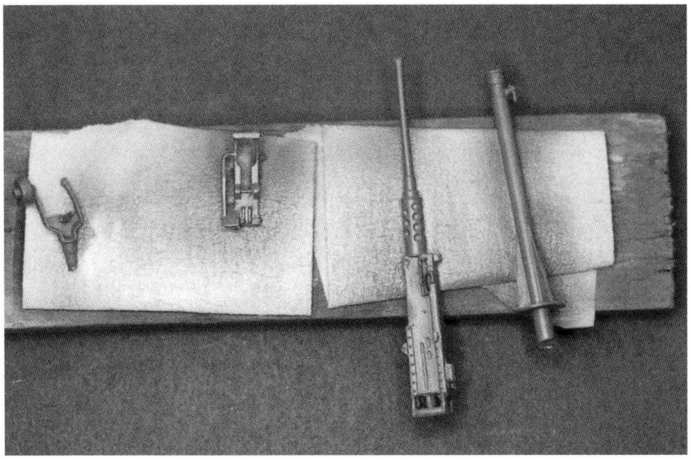

Prime the brass castings to check for flaws and seam lines that you may have missed.

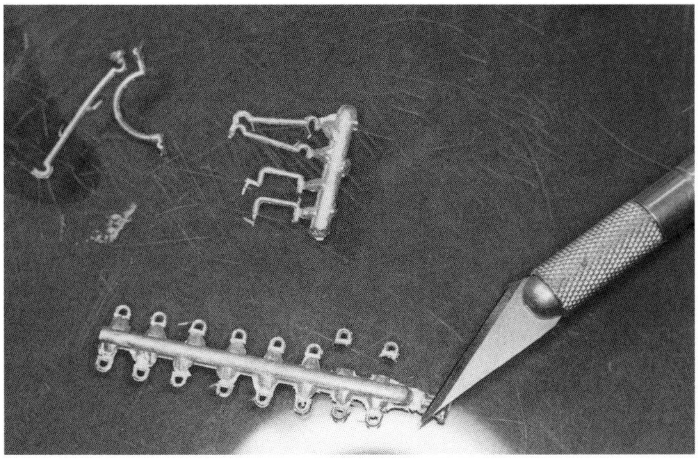

Cut white-metal parts from their trees on a hard surface just as you cut photoetched parts.

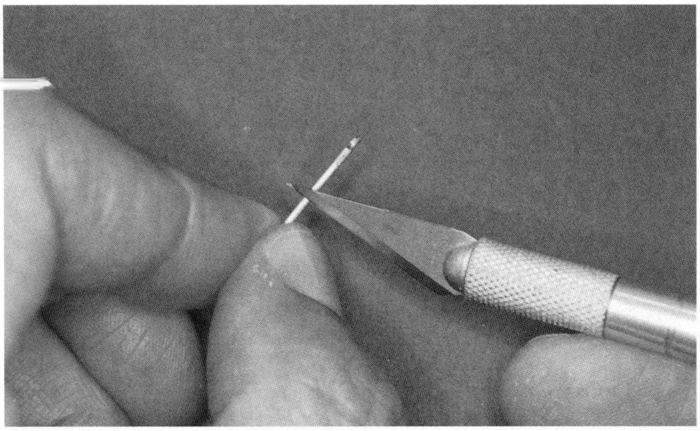

A number 11 X-acto blade works well to scrape off the mold lines on white-metal castings.

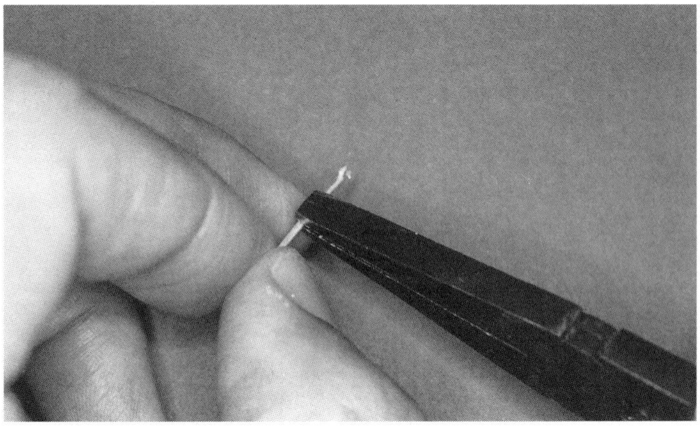

White-metal castings tend to be wavy. It takes a bit of patience with flat-nosed pliers to straighten them out. Work from one end of the casting and straighten out the rod as you progress.